organic at home

Published in 2005 by Murdoch Books Pty Limited.

Murdoch Books Pty Limited Australia
Pier 8/9
23 Hickson Road
Millers Point NSW 2000
Phone: + 61 (0) 2 8220 2000
Fax: + 61 (0) 2 8220 2558

Murdoch Books UK Ltd
Erico House, 6th Floor North
93-99 Upper Richmond Road
Putney, London SW15 2TG
Phone: + 44 (0) 20 8785 5995
Fax: + 44 (0) 20 8785 5985

Chief Executive: Juliet Rogers
Publisher: Kay Scarlett
Creative Director: Marylouise Brammer
Editorial Director: Diana Hill
Editors: Diana Hill, Anouska Jones
Designer: Jacqueline Duncan
Photographers: Alan Benson back cover, and all other pages unless otherwise specified; Marcus Harpur
pages 4 top left, 140-1, 145, 172-3, 183; Greg Delves pages 10-11, 13, 24-5, 52, 56, 62-3, 64, 79;
Lorna Rose page 177; Sue Stubbs front cover, and pages 70-1, 73, 74, 106-7, 109, 115, 117, 154, 169,
170; Jo Whitworth pages 148-9, 158-9, 162-3
Stylists: Justine Brown back cover, and all other pages unless otherwise specified; Jane Campsie pages
10-11, 13, 24-5, 52, 56, 62-3, 64, 79
Writers: Bodyworks section text written by Nerys Purchon; Homeworks section text written by
Alison Haynes; Earthworks section text written by Dr Judyth McLeod, Meredith Kirton, Alison Haynes,
Susan Berry and Steve Bradley
Additional text: Francesca Newby
Production: Monika Paratore

National Library of Australian Cataloguing-in-Publication Data
Organic at Home. Includes index. ISBN 1 74045 372 7. 1. Organic living. 2. Green products.
3. Herbal cosmetics. 4. Organic gardening. 640

Printed by 1010 Printing Limited. PRINTED IN CHINA.

The publisher wishes to thank the following for assistance with photography: The Gooddalls at
Woollybutt, NSW; Ici et Là; MCA Store; Paper Couture; Planet; Planetree Farm; Porter's Paints; Tanya
Laycock; Zoe Macdonell Textile Designs

Notes:
The reader should consult his or her medical, health or other competent professional before adopting
any of the suggestions in this book or drawing inferences from it. The publisher accepts no
responsibility for the effectiveness of the products described herein.
You may wish to perform a patch test for any allergic reactions before applying the bodycare products
described in this book. Put a small quantity of the product on the skin inside your forearm and cover
with a plaster for 24 hours. Any resulting rash or redness could indicate an allergic reaction.

organic
at home

MURDOCH BOOKS

contents

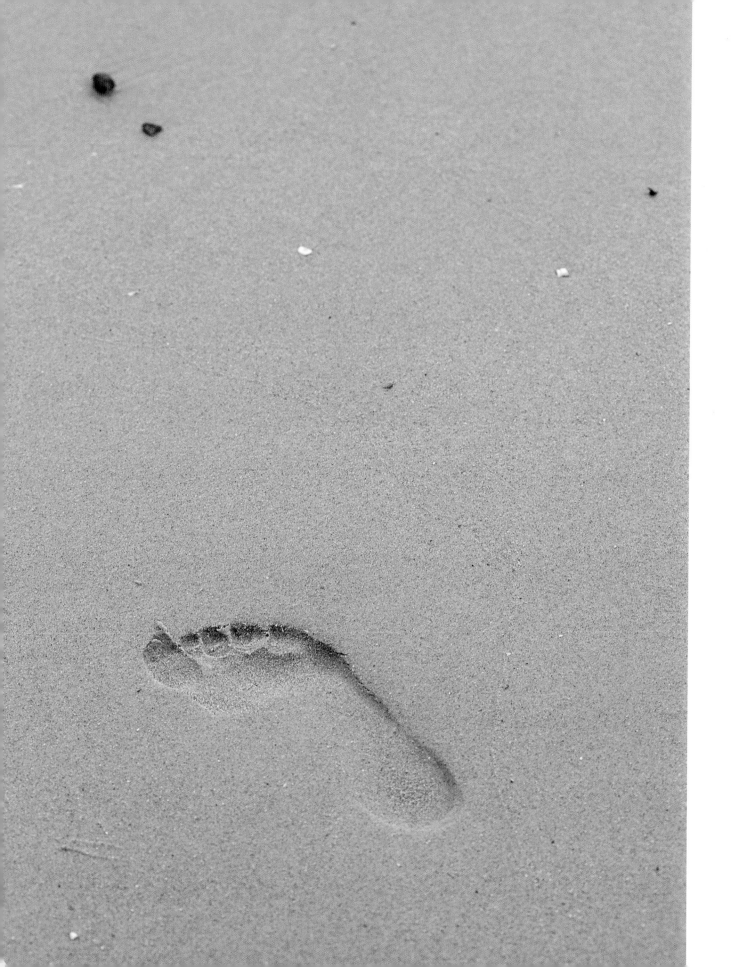

treading lightly

Today there is a growing awareness of the need to tread lightly on our planet in order to protect and heal it — and a realization that signing petitions and recycling our newspapers may not be enough to achieve this. We need to examine our everyday choices and actions to have any hope of making a difference. But getting back to nature is not about unyielding rules and total immersion: it's about adapting your life in a progressive way simply by modifying your routines and habits. You can start as big or as small as you like, bearing in mind that everything you do is a contribution to a larger transformation taking place all around you.

Despite its current prominence in the media, the organic approach suffers from an image problem, often perceived as a way of life that is worthy, but also dull, earnest and spartan. The best-kept secret about the organic phenomenon is that many of the changes you will introduce to your life are addictively luxurious. Whether you are caring for your body, choosing your food, cleaning your house or tending your garden, moving to a more natural way of life is not just about being conscientious, it's also about living well.

In the Bodyworks section we start with the most important person — you. This book is about changing your lifestyle, and the first step towards doing so is changing the way you look after yourself. Going natural is not the same thing as going feral — cleansing and grooming are vital for both physical and mental wellbeing. The key is to make choices that allow you to cherish yourself and protect your surroundings simultaneously. Bodyworks is full of divine recipes for lotions and potions designed for pure pampering. Pleasure is the reward for being good, so start indulging yourself while keeping your principles intact.

The Homeworks section takes your conscience to the heart of your existence, your home. It comes as a shock to realize the nurturing space we have created is in fact a vessel for an array of chemical irritants. 'Anti-bacterial', 'extra-strength', 'power-blast' — the labels of common proprietary household cleaners would have you think your home is a battlefield, as advertizers exploit emotion and guilt to persuade you to buy ever more powerful products. Homeworks is rich in effective grime-busting recipes, proof that gentle is not synonymous with useless.

Earthworks takes the organic approach beyond the walls of the house. Going green in the garden is the final step in the home-grown revolution. It's not only about the joy of eating your own organic vegetables, but also about changing the way you think about your garden. Creating a thriving garden without resorting to the use of chemicals is simple; in this chapter you will find the basic methods that you need to create your own safe green oasis — and, if you want, your own supply of guaranteed non-toxic, highly nutritious edibles.

Whatever your level of commitment, an organic approach will offer you more choices about the way you live. But the first step is often the hardest to take. In order to set out along the natural path you will need some inspiration, information and a plan; in other words, this book.

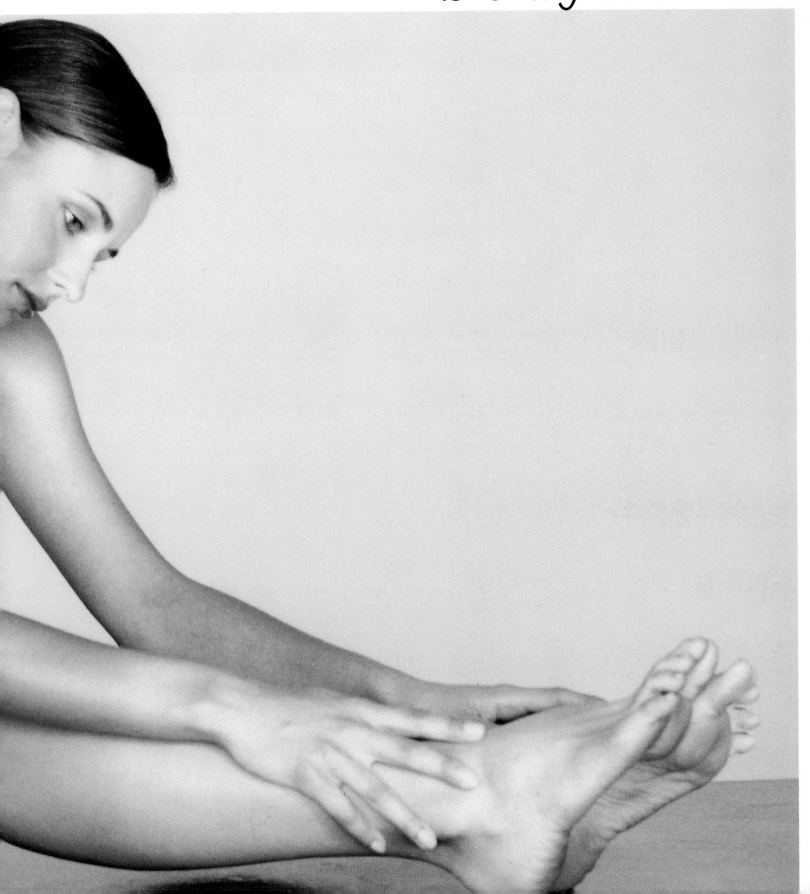

natural body

nourish and protect your skin and body with pure organic ingredients

There is really no excuse for you to neglect any part of your body: cosmetics companies offer a potion or lotion for every nook and cranny. But how much thought do we put into buying them? Do we really buy for results, or are we swayed by the advertizing? How often do we read the list of ingredients inscribed in tiny print on the side of the packaging? And would we still buy so happily if we did?

We strive for the soft, smooth, dewy skin that represents youth, beauty and health. Yet to achieve this we happily smother our skin with products we would never dream of putting in our mouths in case they were somehow toxic. To truly pamper yourself you should give as much thought to your beauty regime as you would to your diet.

The big bonus with preparing your own natural beauty treatments is that it's not difficult and it's not expensive. The making of all the following preparations can take place at the kitchen stove and sink. In fact the recipes are often no more difficult than cooking a meal, and are often much simpler. The less familiar ingredients can be purchased at health food stores and pharmacies.

body care basics

essential oils: use with caution

Keep out of harm's way

Essential oils should be kept out of reach of children, as most are lethal if drunk, even in small quantities.

Essential oils and pregnancy

There are a few oils that are unsafe to use during pregnancy. Check with a doctor before using essential oils.

Oils to use during the first 4 months of pregnancy

Use 1% only in oil blends (1 drop essential oil in 1 teaspoon carrier oil) and 4-5 drops only in the bath: ginger (to sniff if nauseous), grapefruit, mandarin, neroli, spearmint, ylang-ylang.

Oils to use during the last 6 months of pregnancy

Use 1% only in blends (1 drop essential oil in 1 teaspoon carrier oil) and 4-5 drops only in bath: chamomile, grapefruit, geranium and rose geranium, lavender, mandarin, orange, petitgrain, spearmint, ylang-ylang.

sterilizing storage jars

Before starting to prepare your beautycare products, ensure your equipment has been carefully washed in hot, soapy water and the jars you intend to use are thoroughly clean. The best way to ensure that the jars are spotlessly clean is to preheat the oven to very slow 120°C (250°F/Gas 1/2). Thoroughly wash the jars and lids in hot, soapy water (or preferably in a dishwasher) and rinse well with hot water. Put the jars onto baking trays and place them in the oven for 20 minutes, or until you are ready to use them. Dry them fully in the oven. Do not use a teatowel.

Essential oils are highly concentrated, and
when used on the body should be diluted in
a carrier oil such as grapeseed or jojoba.

harvesting and preserving herbs

Harvesting leaves and flowers

Harvest leafy herbs when the concentration of aromatic oils
reaches its highest point, in mid-summer just before
flowering. Leaves should be mature but not show signs of
ageing. After flowering starts, the oils in the leaves are
not as potent. Try to cut whole stems, because you can
handle them without damaging the individual leaves. Only
harvest flowers when they are at their peak, in the middle
of a dry day just before their prime. Do not collect them
when the air is damp or if they are covered in morning
dew, because they will become mouldy. To avoid damaging
the petals, remove whole flowers with some of the stalk,
then discard any damaged ones. Put them in an open
container; in a closed container they may sweat and rot.

Air drying and oven drying

When air drying, remove all lower leaves and wipe off any
moisture on the stems with paper towels. Make bunches of
5 to 10 stems, and secure with an elastic band (if you use
string, the stems may fall through the loop as they dry
and shrink). Hang the bunches upside down in a dark, well
ventilated place at a temperature of about 20°C (68°F)
until they are dry. The drying time will vary from days to
weeks, depending on the thickness of the stems. Store them
in a dark and airtight, labelled glass jar.

You can also dry herbs in a slow conventional oven (no
more than 100°C (200°F/Gas ½)), or even in a microwave
oven. Wrap chopped herbs loosely in paper towel and cook
them for a minute at a time on high. Always have a cup of
water in the microwave, as herbs do not contain much
moisture and the oven could be damaged.

face care

cleansing creams and jellies

Coconut and olive cleanser

Dry/combination/normal skin

This gentle cleanser will keep for a long time in the refrigerator. Refrigeration is essential in warm weather, as otherwise the cream will be very runny.

45 ml (1⅝ fl oz) coconut oil
10 g (¼ oz) cocoa butter
30 ml (1 fl oz) light olive oil
25 ml (⅞ fl oz/1¼ tablespoons)
   macadamia or sweet almond oil
20 drops essential oil of your choice

Melt the coconut oil and cocoa butter together very gently, taking care not to overheat. Stir to mix. Allow to cool a little. Add the olive and macadamia or almond oil and mix well. Cool a little and add the essential oil, mixing thoroughly. Pour into pots.

Coconut honey cleansing cream

Dry/normal skin

Store in the refrigerator, as coconut oil becomes liquid at room temperature.

70 ml (2½ fl oz) coconut oil
20 ml (⅝ fl oz/1 tablespoon)
   light olive or grapeseed oil
10 ml (⅜ fl oz/2 teaspoons) distilled
   water
1½ teaspoons runny honey
10 drops essential oil of your choice

Melt the coconut oil and olive or grapeseed oil in a double boiler until liquid (don't overheat). Remove from the heat. Warm the water and honey to the same temperature as the oil. Slowly drizzle the water and honey into the oils, beating until no drops of water or honey can be seen. Cool slightly and add essential oil, if you want. Beat to emulsify as the mixture cools. Pour into pots.

Honey and peppermint cleansing jelly

Normal/oily/combination skin

1 heaped teaspoon powdered gelatine
1 teaspoon powdered pectin or
   tragacanth
135 ml (4¾ fl oz) hot distilled water
2 teaspoons runny honey
55 ml (1⅞ fl oz) vegetable glycerine
75 ml (2⅝ fl oz) liquid Castile soap
½ teaspoon tincture of benzoin
20 drops peppermint essential oil

Mix the gelatine and pectin or tragacanth together well. Sprinkle over the hot water. Stir until dissolved. Add the remaining ingredients. Bottle while still warm, shaking the jelly occasionally until the mixture is cold.

To use the jelly, massage it gently into wet skin and then rinse off thoroughly. Pat your skin dry.

Make these face care creams in small quantities and store them in the refrigerator for absolute freshness.

Quick milk cleanser

For a good 'quickie' cleanser, use milk (full cream for dry skin, skimmed milk for oily skin) wiped on with a cottonwool ball. There is no need to wash the milk off, as it won't smell and will leave a natural sheen on your skin.

If you would like to add a little refinement to this recipe you can add a handful of elderflower blossoms and a tablespoonful of yoghurt to 125 ml (4 fl oz) milk. Heat to just below boiling and leave covered for half an hour. Strain through a fine sieve and then use.

Milk and honey cleanser

All skin types

This is a quickly stimulating cleanser for all skin types. It removes dead skin, oil and grime.

1 teaspoon dried milk powder
1 teaspoon finely ground almond meal
2 teaspoons runny honey
$1/2$ teaspoon rosewater (optional)
3 drops sweet almond oil

Mix all the ingredients together well.

To use, pat the mixture on from the base of the neck up to the hairline (not around your eyes). Massage gently into the skin. Leave for 10 minutes, then wash off with warm water.

Honey skin cleanser

All skin types

60 ml (2 fl oz/$^1/_4$ cup) runny honey
125 ml (4 fl oz/$^1/_2$ cup) vegetable
    glycerine
2 tablespoons liquid Castile soap

Mix together all the ingredients and pour them into a clean plastic squeezebottle.

To use, dampen the skin and massage a little of the cleanser over the face and throat. Rinse with lukewarm water, then pat your skin dry.

Eye make-up remover

30 ml (1 fl oz) castor oil
30 ml (1 fl oz) light olive oil

Blend both ingredients together.

To use, apply the mixture with a tissue or cottonwool ball to remove make-up from around the eyes. Follow with a regular face cleansing routine.

Milk is an excellent cleanser for fine or dry skin, and can be used to remove other preparations such as face masks.

Washing water

Use this cleansing water to wash your face if your skin is very sensitive, or if you are troubled with acne, pimples or other skin problems. This water may also be used as a moisturizing lotion with the addition of 1 teaspoon of sweet almond, macadamia or any other fine-textured oil.

40 ml ($1^3/_8$ fl oz/2 tablespoons) rosewater
60 ml (2 fl oz/$^1/_4$ cup) distilled water
$^1/_4$-$^1/_2$ teaspoon vegetable glycerine (Try
  $^1/_4$ teaspoon first, and increase
  quantity to $^1/_2$ teaspoon if you wish)
6 drops essential oil of your choice

Mix all ingredients together in a 100 ml ($3^1/_2$ fl oz) bottle and shake well to blend thoroughly. Leave for 4 days to synergize, shaking occasionally. Filter the mixture through coffee filter paper. Store in a dark-coloured glass bottle and shake well before use.

To use, take a palm-sized piece of cottonwool, dip it in warm water and squeeze it out well. Flatten it out into a pad, sprinkle with Washing water and use to cleanse the throat and face. Repeat if necessary. There is no need to use a toner after this treatment.

Milk and honey soap

If you have a normal, combination or oily skin type, you may prefer to use soap to cleanse your face and neck. The glycerine and honey are both humectants, which means that they attract moisture to the skin and hold it there.

375 g (13 oz/$4^1/_2$ cups) finely grated
  unscented pure soap or soap scraps
20-30 ml ($^5/_8$-1 fl oz/
  1-$1^1/_2$ tablespoons) milk
2 teaspoons vegetable glycerine
2 teaspoons runny honey
20 drops rosemary essential oil
10 drops lavender essential oil
10 drops peppermint essential oil

Put the grated soap and milk in a bowl over a pan of simmering water, stirring occasionally until melted. If the mixture is too thick, add more milk and reheat. Use the least amount of milk possible, or the soap will shrink a lot as it dries. Add the glycerine and honey and stir until completely incorporated. Add the essential oils when the mixture has cooled slightly.

Press the soap into moulds (such as soap dishes, individual jelly moulds, milk cartons cut in half, little baskets lined with muslin), or shape it into balls. Set aside in an airy place to dry for 2 to 6 weeks.

When the soap has hardened, you might like to polish it. Lightly moisten a cloth with some water to which a drop or two of essential oil has been added, and then buff the soap.

Cucumber has both
antiseptic and anti-
inflammatory properties.

toners and astringents

Skin toners restore the acid mantle to your skin and leave it feeling fresh and clean. These preparations are also suitable for pre- and after-shave lotions.

To apply, wet some cottonwool with water and squeeze it dry (this prevents any wastage of your precious lotion). Sprinkle the cottonwool with a few drops of the lotion and stroke it upwards over your throat and face.

Rosewater and witch-hazel toner

All skin types

The following is a simple and effective toner for all skin types. If your skin is dry you can decrease the distilled witch-hazel and increase the rosewater. If you have oily skin you can do the reverse.

100 ml (3$^{1}/_{2}$ fl oz) rosewater or hydrosol
30 ml (1 fl oz) distilled witch-hazel
$^{1}/_{2}$ teaspoon vegetable glycerine

Mix all the ingredients together. Bottle them.

Calendula skin toner

All skin types

This is a toner for all skin types, particularly those troubled by spots or blemishes. The toner will keep in the refrigerator for up to 3 weeks or it can be frozen in ice-cube trays and defrosted as needed.

135 ml (4$^{3}/_{4}$ fl oz) boiling distilled water
2 teaspoons dried, crushed calendula
    flowers
20 drops tincture of benzoin
10 drops essential oil of your choice
    (optional)

Pour the boiling distilled water over the dried calendula flowers, cover and leave until cold. Strain through a sieve. Add the remaining ingredients, mix well and leave to stand, covered, for 6 to 8 hours. Strain through coffee filter paper. Bottle or pour into ice-cube trays.

Elderflower water

This is a soothing, healing, lotion for any kind of skin that has been in the wind and sun and is feeling dry and rough. It will keep for 2 to 3 weeks in the refrigerator but may also be frozen in ice-cube trays and defrosted as needed.

135 ml (4³/₄ fl oz) boiling distilled water
2 teaspoons dried elderflower blossooms
1 teaspoon vegetable glycerine
20 drops tincture of benzoin

Pour the boiling distilled water over the dried elderflowers, cover and allow to cool. Strain through a sieve. Add the remaining ingredients, mix well and then leave to stand, covered, for 6 to 8 hours. Strain the mixture through coffee filter paper. Bottle.

Cucumber toner

Cucumber toner doesn't keep well unless preserved in some manner. If you want to store it for a short time, say 7 to 21 days, use tincture of benzoin as a mild preservative. The toner may also be frozen in ice-cube trays and defrosted as needed.

125 ml (4 fl oz/½ cup) cucumber juice
125 ml (4 fl oz/½ cup) distilled
  witch-hazel
1 teaspoon tincture of benzoin

Mix all the ingredients together. Strain through a fine sieve and then through coffee filter paper. Bottle.

facial steaming

Facial steaming causes the skin to perspire, which helps to deep-cleanse every pore; it also loosens grime and dead skin cells. The heat from the steam stimulates the blood supply and hydrates the skin, which will look and feel softer and more youthful.

If you have 'thread veins' on the cheeks, you need to be careful when steaming. Apply a thick layer of moisture or night cream over the veins and hold your face about 40 cm (15 inches) away from the steam, no closer. Don't steam more than once a fortnight. If your skin is dry and sensitive, apply a thin layer of honey on the face and throat before you steam. Keep your face 40 cm (15 inches) away from the steam. Don't steam more than once a fortnight.

Normal, combination and oily skins may be steamed as often as twice a week.

Have ready a shower cap, a large towel and a heatproof pad for the table.

Wash or otherwise cleanse the face. Don the shower cap.

Put 4 tablespoons finely chopped herbs in a pan. Pour over 2 litres (70 fl oz) cold water and cover the pan with a lid.

Bring to the boil, reduce the heat and simmer gently for 5 minutes, then put the pan on the heatproof mat and remove the lid. Add the essential oils to the pan.

Form a tent with the towel over the pan and your head. Those with problem-free skin can keep their face 20 cm (8 inches) away from the steam, for 5 to 10 minutes. Deep grime will be drawn to the surface of the skin.

Splash your face with cool (not cold) water and finish with a tonic or astringent and some moisturizer.

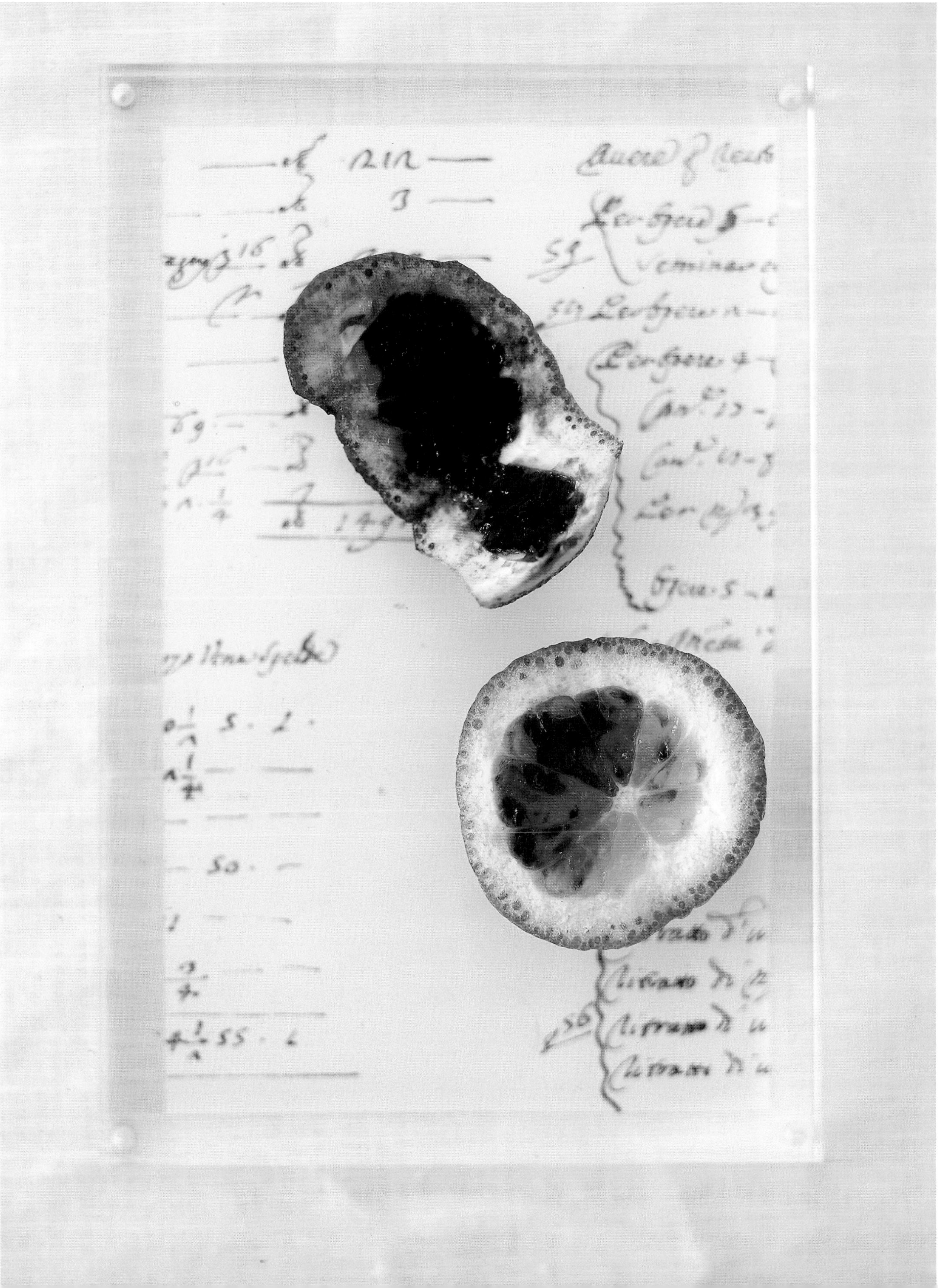

a note on ingredients

Herbs and citrus peel should be dried and finely chopped or rubbed. Seeds should be dried and crushed, and roots dried and ground or finely chopped. Essential oils may be used alone or with dried or fresh herbs. Add the oils immediately before using the treatment. No more than 8 drops of essential oil may be used in 2 litres (70 fl oz) boiling water. To prepare these recipes, see the instructions on page 26.

Herbal blend for dry skin

5 g (1/8 oz/1/4 cup) dried chamomile flowers and leaves

15 g (1/2 oz/1 cup) dried clover blossoms

25 g (1 oz/1/4 cup) dried ground comfrey root (comfrey is a restricted substance in some countries)

35 g (1 1/4 oz/1 cup) dried comfrey leaves (see note for root)

60 g (2 1/4 oz/1 cup) dried fennel seeds and leaves

25 g (1 oz/1 cup) dried violet leaves and flowers

Herbal blend for oily skin

30 g (1 oz/1 cup) dried lemongrass

45 g (1 1/2 oz/ 1 cup) dried lemon peel and leaves

25 g (1 oz/1/4 cup) dried liquorice root

80 g (2 3/4 oz/1 cup) dried comfrey root and leaves (comfrey is a restricted substance in some countries)

5 g (1/8 oz/1/4 cup) dried peppermint leaves

Herbal blend for normal/combination skin

60 g (2 1/4 oz/1 cup) dried fennel seeds and leaves

20 g (3/4 oz/1 cup) dried lemon peel and leaves

20 g (3/4 oz/1 cup) dried orange peel and leaves

20 g (3/4 oz/1 cup) dried lavender flowers and leaves

Blend for combination skin

70 ml (2 1/2 fl oz) sweet almond oil
8 drops geranium essential oil
6 drops palmarosa essential oil
6 drops lavender essential oil

Mix all the ingredients together in a small bottle. Leave for 4 days to blend.

Essential oil blend for dry skin

70 ml (2 1/2 fl oz) sweet almond oil
8 drops rosewood essential oil
8 drops palmarosa essential oil
4 drops rose or lavender essential oil

Mix all the ingredients together in a small bottle. Leave for 4 days to blend.

Essential oil blend for oily skin

75 ml (2 5/8 fl oz) grapeseed oil
12 drops lemon essential oil
4 drops patchouli essential oil
4 drops sandalwood essential oil

Mix all the ingredients together in a small bottle. Leave for 4 days to blend.

You can dry your own citrus peel on wire cooling racks in a warm, well-aired position.

scrubs and masks

Scrubs and masks may be used by people of either sex and all ages. They may be used for exfoliating, clearing excessive oiliness, refining pores, nourishing dry skin and improving circulation.

The frequency with which you use scrubs and masks depends entirely on your skin type. If you have oily blemished skin you will be able to use these preparations several times a week, but if your skin is fine and dry it would be wise to choose only the most gentle treatment and use it perhaps once a fortnight. Areas with obvious 'thread veins' should never be treated with masks or scrubs as the additional stimulation could cause a worsening of the condition.

Scrubs may be made in bulk, stored in a glass jar in the bathroom and mixed as needed. However, as most masks are made with fresh fruit, vegetables and milk or yoghurt, they should be prepared as needed or kept for no longer than 24 hours in the refrigerator.

Almond scrub

All skin types

Almonds are very softening to the skin and this recipe is gentle enough to use as a regular cleanser provided that the almonds are very finely ground.

2 teaspoons sweet almond oil
4 teaspoons very finely ground almonds
1 teaspoon cider vinegar
1 drop lavender or palmarosa essential oil
distilled water

Mix all the ingredients to a smooth paste, adding distilled water as needed. To use, massage gently into the skin, rinse off with lukewarm water and pat your skin dry.

Parsley and lettuce scrub

All skin types

Lettuce and parsley are good for the skin.

2 parts powdered bran
2 parts powdered oatmeal
2 parts powdered soap
1 part dried parsley, powdered
1 part dried lettuce, powdered
1 part dried comfrey leaf, powdered
   (comfrey is a restricted substance in
   some countries)

Mix all the ingredients together. Store in an airtight jar. To use, put 4 teaspoons in a small bowl. Add enough water or milk to form a soft paste and massage gently into the skin using small circular movements. Rinse off with cool water. Blot your skin dry with a soft towel.

Yoghurt and yeast scrub

Normal/combination/oily skin

Yeast stimulates the circulation, bringing blood to the surface of the skin. Be very careful when using this scrub not to overstimulate the cheeks, where the delicate capillary veins lie near the surface. This recipe is sufficient for one treatment. It does not keep.

4 teaspoons yoghurt
2 teaspoons almond meal
1 teaspoon brewer's yeast
1 teaspoon runny honey

Mix all the ingredients together. Use the scrub immediately. To use, gently massage onto the skin. Rinse off with lukewarm water, then pat your skin dry.

The finely ground almonds and oatmeal used in these scrubs thicken the preparations as well as refine the skin.

Egg white helps to tighten and
smooth the skin.

Basic mask

Make a basic mask in advance and store it
in an airtight jar ready for mixing.

100 g (3½ oz) white clay
25 g (1 oz) cornflour
4 teaspoons very finely ground oats
4 teaspoons very finely ground almond meal

Mix all the ingredients well and store in
a tightly covered jar.

To use, mix 4 teaspoons of the basic mask
into a soft paste with honey, fruit juice
or pulp, vinegar, egg, oil or herbal tea.
Spread the mask over your face and neck
(be very careful if you have dry skin or
'broken' veins, as masks may be over-
stimulating). Lie down on your bed or in
the bath and relax for 15 to 20 minutes.
You could also place cucumber slices or
cottonwool soaked in distilled witch-hazel
on your eyes. Wash the mask off in warm
water and follow with a cool splash.

Quick honey mask

A simple honey mask is a stimulating
cleanser for all skin types. It can be
done very quickly, easily and cheaply.

Spread some warm runny honey over your
face and very gently begin to tap your
skin with your fingertips until there is
a feeling of 'pulling'. This takes about
2 minutes. Stop tapping and rinse your
face well with coolish water. Blot dry.

Egg and lemon mask

This mask may be used as a cleanser if
the wholemeal (wholewheat) flour is
replaced with arrowroot. It is a deep
cleanser that leaves skin feeling soft. If
your skin is dry, replace the lemon juice
with orange juice for a gentler action.

1 egg white, beaten
1 teaspoon olive or sweet almond oil
2 teaspoons lemon or orange juice
4 drops lemon or orange essential oil
wholemeal flour, to thicken

Mix the ingredients together and store in
the refrigerator. Use within 3 days.

To use, spread the mask over your face and
neck (be very careful if you have dry skin
or 'broken' veins, as masks may be over-
stimulating). Lie down and relax for 15 to
20 minutes. Wash the mask off in warm
water and follow with a cool splash.

Quick egg mask

An egg white lightly beaten, smoothed on
the skin and rinsed off after 20 minutes,
will tighten and tone your skin. If your
skin is dry you should smear some honey
or oil on your skin before spreading on
the egg white. If you have oily skin,
½ teaspoon lemon juice may be beaten
into the egg white.

moisturizers

satin smooth moisture lotion

Most skin types

This is a lovely soft lotion that is so moisturizing it can double as body oil. Store in the refrigerator and shake well before use.

Phase A
150 ml (5 fl oz) distilled water
$1/2$ teaspoon borax
50 drops grapefruit seed extract (optional preservative)

Phase B
$1/2$ teaspoon anhydrous lanolin
3 teaspoons finely grated beeswax
60 ml (2 fl oz/$1/4$ cup) sweet almond oil
40 ml ($1^3/8$ fl oz) light sesame or jojoba oil

Phase C
10 drops essential oil of your choice

Gently melt all the ingredients for Phase B together in a double boiler. Stir until all the ingredients are melted but not overheated.

Combine all ingredients for Phase A and heat until the mixture is the same temperature as Phase B.

Slowly trickle Phase A into Phase B, stirring constantly and reheating gently if the mixture begins to thicken.

Remove from the heat. Stir until the outside of the container feels just a little hotter than your hand.

Add the essential oil from Phase C and mix thoroughly.

Spoon into sterilized containers.

Beeswax not only moisturizes, it also helps to emulsify and thicken skincare preparations.

Almond rose cream

Dry/combination/normal skin

This moisturizing cream should be stored in the refrigerator.

Phase A
60 ml (2 fl oz/$^1/_4$ cup) rosewater
1 teaspoon borax
40 drops grapefruit seed extract (optional preservative)

Phase B
90 ml (3 fl oz) sweet almond oil
90 ml (3 fl oz) jojoba or macadamia oil
2 teaspoons finely grated, tightly packed beeswax
1000 IU vitamin E d'alpha tocopherol

Phase C
20 drops essential oil of your choice

Gently melt the ingredients for Phase B together in a double boiler (prick the vitamin capsules and squeeze out the contents). Stir until all are melted but not overheated. Combine the ingredients for Phase A and heat until the mixture is the same temperature as Phase B. Slowly trickle Phase A into Phase B, stirring constantly and reheating gently if the mixture begins to thicken. Remove from the heat. Stir until the outside of the container feels just a little hotter than your hand. Add the essential oil that comprises Phase C and mix thoroughly. Spoon the almond rose cream into sterilized containers.

Luxurious moisturizer

Dry/normal/damaged skin

Phase A

1 teaspoon vegetable glycerine
40 ml (1³/₈ fl oz) distilled water
1 teaspoon borax
20 drops grapefruit seed extract (optional
  preservative)

Phase B

2 teaspoons finely grated, tightly packed
  beeswax
2 teaspoons finely grated, tightly packed
  cocoa butter
¹/₂ teaspoon liquid lecithin
2 teaspoons avocado oil
2 teaspoons light olive oil
4 teaspoons sweet almond oil
4 teaspoons apricot oil
1000 IU vitamin E d'alpha tocopherol

Phase C

20 drops essential oil of your choice
10 drops carrot seed oil

Gently melt the ingredients for Phase B together in a double boiler (prick the vitamin capsules and squeeze out the contents). Stir until all are melted but not overheated. Combine ingredients for Phase A and heat until the mixture is the same temperature as Phase B. Slowly trickle Phase A into Phase B, stirring constantly and reheating gently if the mixture begins to thicken. Remove from the heat. Stir until the outside of the container feels a little hotter than your hand. Add Phase C oils and mix thoroughly. Spoon into sterilized containers. Store in the refrigerator.

Honey and vitamin rich night cream

Dry skin

Phase A

80 ml (2¹/₂ fl oz/¹/₃ cup) warm
  distilled water
1 teaspoon runny honey

Phase B

40 g (1¹/₂ oz) lanolin
1000 IU vitamin E d'alpha tocopherol
¹/₂ teaspoon sweet almond oil
5 drops carrot seed oil
¹/₂ teaspoon liquid lecithin

Phase C

6 drops lavender essential oil
2 drops patchouli essential oil

Gently melt the ingredients for Phase B together in a double boiler (prick the vitamin capsules and squeeze out the contents). Stir until all are melted but not overheated. Combine Phase A ingredients and heat until the honey is melted and the temperature of the mixture is the same as Phase B. Slowly trickle Phase A into Phase B, stirring constantly and reheating gently if the mixture begins to thicken. Remove from the heat. Stir until the outside of the container feels just a little hotter than your hand. Add the Phase C oils and mix thoroughly. Spoon the cream into sterilized containers. Store in the refrigerator.

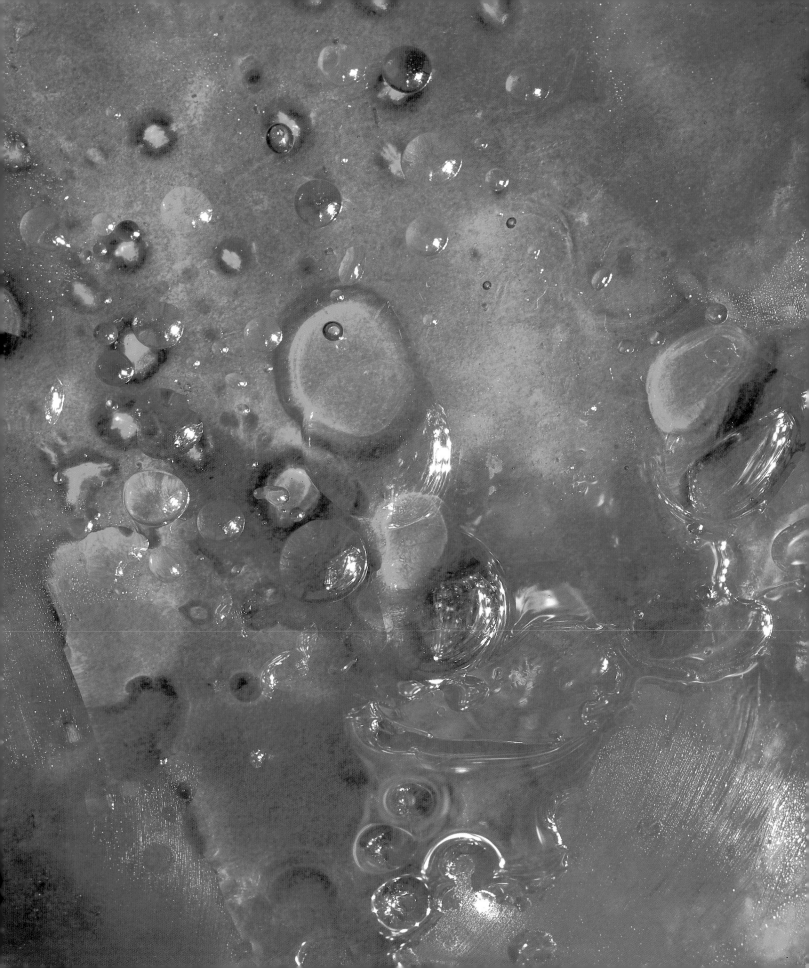

If you have sensitive skin, test a tiny amount of oil on the skin of your forearm to check for an allergic reaction.

Regenerating oil for dry skin

80 ml (2¹/₂ fl oz/¹/₃ cup) macadamia oil
1 teaspoon avocado oil
40 drops evening primrose oil
40 drops jojoba oil
1 teaspoon rosehip oil
15 drops palmarosa essential oil
5 drops lavender essential oil
5 drops sandalwood essential oil
5 drops ylang-ylang essential oil

Combine ingredients in a 100 ml (3¹/₂ fl oz) bottle. Shake well for several minutes. Leave for 4 days to blend; store in a cool dark place. Shake the bottle well before use. Gently massage the oil into slightly dampened skin. Use morning and night.

Regenerating oil for normal skin

80 ml (2¹/₂ fl oz/¹/₃ cup) sweet almond oil
1 teaspoon avocado oil
1 teaspoon wheatgerm oil
1 teaspoon jojoba oil
5 drops lavender essential oil
15 drops palmarosa essential oil
5 drops rosewood essential oil
5 drops sandalwood essential oil

Combine all the ingredients in a 100 ml (3¹/₂ fl oz) bottle. Shake well for several minutes. Leave for 4 days to blend; store in a cool dark place. Shake well before use. Gently massage the oil into slightly dampened skin.

Regenerating oil for oily skin

Those with oily skin will have noticed that there are areas on the face and neck which have little or no oil: these are mainly the throat, lips and under the eyes. Use this treatment on these areas but also use a thin smear over the whole face before going to bed, as the oils have the capacity to 'balance' and moisturize the skin without encouraging the oil glands to produce more sebum.

80 ml (2¹/₂ fl oz/¹/₃ cup) grapeseed oil
2 teaspoons apricot oil
40 drops evening primrose oil
10 drops bergamot essential oil
5 drops lemon essential oil
5 drops lavender essential oil
10 drops sandalwood essential oil

Combine all the ingredients in a 100 ml (3¹/₂ fl oz) bottle. Shake well for several minutes. Leave for 4 days to blend; store in a cool dark place. Shake the bottle well before use. Gently massage the oil into slightly dampened skin. Use morning and night.

Regenerating oil for the neck

The skin on the neck contains very few oil glands, so it needs a little more attention. This oil blend is very powerful and should be used at night to get the best result.

2 teaspoons avocado oil
2 teaspoons evening primrose oil
55 ml ($1^7/_8$ fl oz) jojoba oil
4 teaspoons hazelnut oil
20 drops carrot seed oil
15 drops palmarosa essential oil
5 drops rosewood essential oil
5 drops frankincense essential oil
5 drops geranium essential oil

Combine all ingredients in a 100 ml ($3^1/_2$ fl oz) bottle. Shake well for several minutes. Leave for 4 days to blend. Store in a cool dark place.

To use, shake the bottle. Measure 5 to 10 drops into the palm of your hand and, with a light, upward motion, use the fingers of the other hand to spread the oil from the collarbones to the chin. Massage in, leave for 20 minutes, then blot the surplus off with a tissue.

Around-the-eyes oil

The thin, fine skin around the eyes shows early lines in the same way as the neck. Remember that the skin around the eyes is super fragile and you can do more harm than good if you are heavy-handed. Use only the middle finger to gently pat creams and lotions on this area and avoid using heavy oils that can 'drag'.

50 ml ($1^3/_4$ fl oz) hazelnut oil
15 drops jojoba oil
35 drops evening primrose oil
15 drops carrot seed oil
1000 IU vitamin E d'alpha tocopherol

Place all the ingredients in a 50 ml ($1^3/_4$ fl oz) bottle (prick the vitamin capsules and squeeze the contents into the bottle). Shake the bottle very well to mix. Leave for 4 days to blend. Store in a cool dark place.

To use, shake the mixture well. One drop under each eye should be enough. Apply the oil at night, leave it for 10 minutes and then carefully blot the excess off with a tissue. Avoid getting the oil into the eye itself or it will sting and could be harmful.

When you buy essential oils, check the label to see where the manufacturer sources its ingredients, and if it runs checks for purity.

body care

moisturizers

Body butter

1 teaspoon finely grated, tightly packed
  beeswax
4 teaspoons cocoa butter
1 teaspoon coconut oil
1 teaspoon shea butter, chopped
2 teaspoons cold-pressed avocado oil
2 tablespoons sweet almond oil
1000 IU vitamin E d'alpha tocopherol
4 drops ylang-ylang essential oil
2 drops sandalwood essential oil

Heat the beeswax and cocoa butter in the
top half of a double boiler until melted.
Remove from the heat and add the coconut
oil, stirring until it has melted. Cool to
just above hand heat (gauged when you feel
the outside of the pan). Add the shea
butter, avocado oil and almond oil,
beating with a whisk until smooth. Prick
the vitamin capsules and squeeze the
contents into the mixture. Add the
essential oils a drop at a time, beating
after each addition until thoroughly
incorporated. Spoon into sterilized jars.

Moisture blocks

80 g (2³/₄ oz) cocoa butter
1 teaspoon coconut oil
1 teaspoon avocado oil
1 teaspoon sweet almond oil
1 teaspoon light olive oil
1 teaspoon jojoba oil
1000 IU vitamin E d'alpha tocopherol
10 drops lavender essential oil
5 drops patchouli or sandalwood
  essential oil

Melt the cocoa butter in the top of a
double boiler. When melted, remove from
the heat and add all the remaining
ingredients, except for the essential oils
(prick the vitamin capsules and squeeze
out the contents). Cool the mixture
slightly, then add the essential oils and
mix thoroughly. Pour into little chocolate
moulds or mini muffin tins. Place in the
refrigerator to cool.

Moisturizing blocks can be used as an
after-shower treatment for the body while
the skin is still very slightly damp.

aromatherapy massage oils

Basic massage oil

80 ml (2^1/$_2$ fl oz) fractionated coconut
    oil or macadamia oil
2 teaspoons sweet almond oil
2 teaspoons avocado oil
1000 IU vitamin E d'alpha tocopherol

Mix ingredients together in a 100 ml
(3^1/$_2$ fl oz) bottle (prick the vitamin
capsules and squeeze out the contents).
Shake well to mix.

Aching bodies

This is a good blend for an after-sport
massage when you ache all over.

20 drops lavender essential oil
10 drops rosemary essential oil
10 drops clary sage essential oil
5 drops peppermint essential oil
5 drops cypress essential oil

Add the essential oils to an entire bottle
of Basic massage oil. Shake well to mix.

Aching minds

For when you are tired from too much
thinking and worrying.

10 drops lavender essential oil
20 drops chamomile essential oil
10 drops geranium essential oil
5 drops cedarwood essential oil
5 drops peppermint essential oil

Add the essential oils to an entire bottle
of Basic massage oil. Shake well to mix.

Sleep aid

Use this blend for a gentle massage that
will relieve stress and ensure a good
night's sleep.

10 drops chamomile essential oil
20 drops lavender essential oil
10 drops marjoram essential oil
10 drops sandalwood essential oil

Add the essential oils to an entire bottle
of Basic massage oil. Shake well to mix.

Massage strokes should work in
harmony with the body's blood flow —
always massage towards the heart.

hand care

Hand cleanser

This simple hand cleanser is very thorough but still gentle.

50 g (1³/₄ oz) finely powdered soap
50 g (1³/₄ oz) fine sawdust
¹/₂ teaspoon borax

Mix all the ingredients together. Store in an airtight jar.

To use, put a teaspoonful of the mixture on the palm of your hand, drizzle on enough water to form a paste and work this up to a lather. Rinse your hands and pat them dry.

Quick lemon and sugar scrub

1 tablespoon lemon juice
1-2 tablespoons sugar

Mix both the ingredients together. Use the scrub immediately.

To use, massage the mixture into your hands, rinse off, dry and then apply a little of the Lemon and almond hand cream. Your hands will feel like smooth satin.

Lemon and almond hand cream

Phase A
30 ml (1 fl oz) vegetable glycerine
20 ml (⁵/₈ fl oz/1 tablespoon)
    distilled water
1 teaspoon borax

Phase B
22 g (³/₄ oz) grated beeswax
60 ml (2 fl oz/¹/₄ cup) sweet almond oil
1 teaspoon castor oil
60 ml (2 fl oz/¹/₄ cup) olive oil

Phase C
20 drops lemon essential oil
20 drops lavender essential oil

Gently heat all Phase A ingredients until the borax is dissolved. Melt the Phase B ingredients gently in a double boiler until the mixture is liquid but not overheated. Slowly trickle the Phase A mixture into the Phase B mixture. Stir constantly and reheat slightly if the mixture begins to solidify. Add the Phase C ingredients when the outside of the container feels a little hotter than your hand and the mixture is still liquid. Stir until the essential oils are completely incorporated. Pour into sterilized containers and cap at once.

Sweet almond oil is an excellent treatment for dry and chafed skin, while glycerine will prevent moisture evaporation.

Rosewater and glycerine hand lotion

The vinegar in this recipe restores the acid balance of the skin and helps the other ingredients to be easily absorbed into the skin.

20 ml (^{5}/$_8$ fl oz/1 tablespoon) rosewater
2 teaspoons vegetable glycerine
^{1}/$_2$ teaspoon white wine vinegar
^{1}/$_2$ teaspoon runny honey
10 drops lemon essential oil

Mix all the ingredients together in a 50 ml (1^3/$_4$ fl oz) bottle. Shake well.

Honey hand lotion

This simple hand lotion is easy to make and use. It will separate on standing, so it should be shaken before use.

1 tablespoon runny honey
1 teaspoon avocado oil
1 teaspoon sweet almond oil
80 ml (2^1/$_2$ fl oz/^{1}/$_3$ cup) rosewater
1 teaspoon vegetable glycerine
1 tablespoon white wine vinegar

Combine the honey and oils and heat gently until the honey is melted. Heat the remaining ingredients until they are roughly the same temperature as the oils and honey. Add to the oil and honey mixture and beat together until cool. Bottle and shake well.

Hands are at the frontline of exposure to harsh and drying household cleaning agents; repair them with an intensive essential oil treatment.

Hand and nail essential oil

Essential oils are particularly good for hands and nails as they work very quickly and are readily absorbed without leaving an unpleasant greasy feeling.

45 ml (1^5/$_8$ fl oz) macadamia oil
1 teaspoon avocado oil
20 drops carrot seed oil
30 drops jojoba oil
40 drops evening primrose oil
15 drops lemon essential oil
10 drops rosemary essential oil
10 drops geranium essential oil
5 drops patchouli essential oil
1000 IU vitamin E d'alpha tocopherol

Mix all the ingredients together in a 50 ml (1^3/$_4$ fl oz) bottle (prick the vitamin capsules and squeeze out the contents). Shake well to mix, then leave for 4 days to blend. Store in a cool dark place.

To use, shake the bottle and then massage 4 to 5 drops into the hands and around the nail bed.

foot care

Herbal foot bath

To make this soothing bath for sore feet, first brew a strong tea of finely chopped fresh herbs to half fill a bowl large enough for both feet. Use as many of these herbs as possible: lavender, sage, pine needles, pennyroyal, rosemary and yarrow.

Have the herbal water as hot as you can bear. Soak your feet for 10 to 15 minutes. Finish by plunging your feet into a bowl of cold water mixed with 4 tablespoons of witch-hazel. Pat your feet dry and massage with Aromatherapy foot oil.

Foot powder

This powder may be sprinkled inside socks and shoes if you suffer from sweaty feet.

100 g ($3^1/_2$ oz/1 cup) kaolin powder
65 g ($2^1/_4$ oz/$^1/_2$ cup) cornflour
55 g (2 oz/$^1/_2$ cup) arrowroot
2 tablespoons powdered dried sage leaves
2 tablespoons powdered dried rosemary
10 drops lavender essential oil
10 drops tea tree essential oil
10 drops tincture of myrrh
10 drops tincture of benzoin

Mix all the dry ingredients together, and blend or rub until very fine. Push through a fine sieve. Mix the oils and tinctures together and drip slowly onto the mixed powders, stirring constantly to avoid clumping. Store in a container with a very tight-fitting lid.

Aromatherapy foot oil

Your feet will thank you for massaging them with this beautiful oil. Not only does it ease the pain of tired feet but it also soothes the irritating symptoms of tinea.

4 teaspoons vodka
45 ml ($1^5/_8$ fl oz) light olive oil
2 teaspoons avocado oil
2 teaspoons jojoba oil
2 teaspoons sweet almond oil
20 drops peppermint essential oil
10 drops tea tree essential oil
30 drops lavender essential oil
20 drops cypress essential oil

Put all the ingredients into a 100 ml ($3^1/_2$ fl oz) bottle. Shake well for several minutes. Leave for 4 days to blend. Store in a cool dark place.

To use, first wash or soak your feet, then dry them thoroughly. Shake the bottle well and massage a little of the oil into your feet until it is absorbed.

To massage your feet, apply your chosen oil and knead the sole of each foot. Then slowly work over the top of the foot and up the leg with a long, sweeping action.

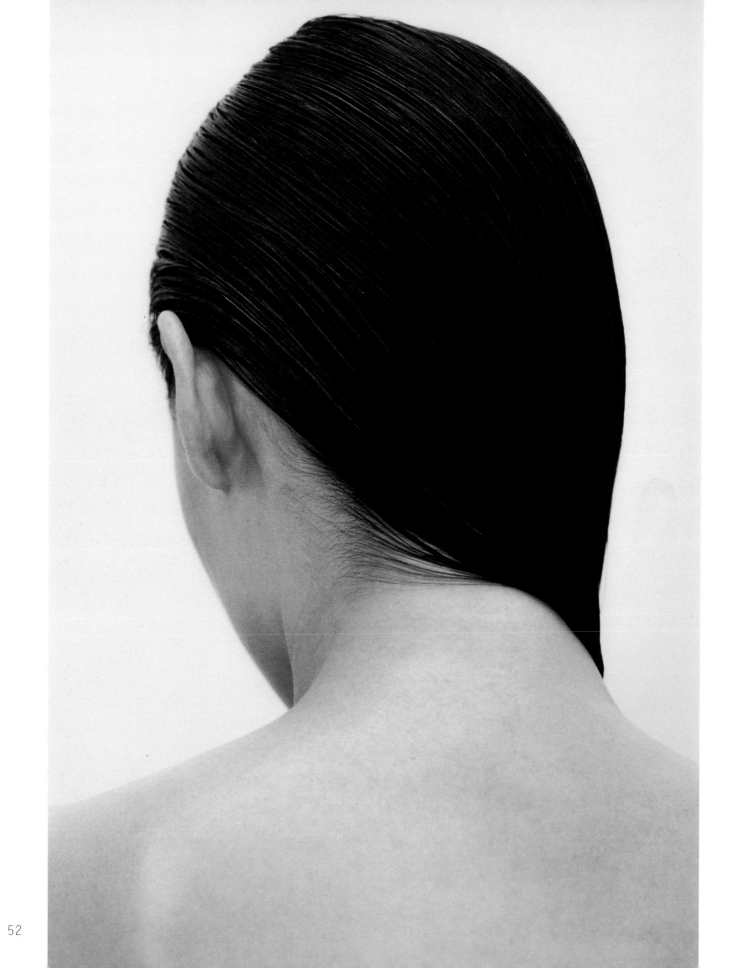

hair care

pre-shampoo treatments

Simple egg treatment

All hair types

This is a really effective protein treatment for hair. Beat 2 eggs in a bowl until light and frothy. Massage this mixture thoroughly into your hair. Rinse with lukewarm water. If you rinse with hot water you'll end up with scrambled egg in your hair!

Hair mayonnaise

All except very oily hair

1 egg yolk
3 teaspoons apple cider vinegar
4 teaspoons castor oil
4 teaspoons light olive oil
4 drops lavender essential oil

Beat the egg and vinegar together with a wire whisk or in a blender. Mix the oils together and add to the egg and vinegar mixture in a thin stream until the mixture is thick. Use immediately.

To use, massage the mayonnaise into your hair, then cover your hair with a plastic shower cap before wrapping a hot towel around your head. Leave for 20 minutes, and then wash your hair using a mild herbal shampoo.

Honey and lemon treatment

Oily hair

1 egg white, beaten
1 teaspoon runny honey
1 teaspoon lemon juice
4 teaspoons water
3 drops lemon essential oil
dried milk powder

Combine all the ingredients, except for the dried milk powder, together in a small bowl. Add enough milk powder to make a smooth, soft paste. Use immediately.

To use, massage the paste into your hair. Cover your hair with a plastic shower cap before wrapping a hot towel around your head. Leave for 20 minutes, then shampoo your hair with a mild herbal shampoo.

The condition of your hair reflects your inner health. The way to strong, shining hair is via a nutritious, balanced diet and an effective health and fitness regime.

shampoos and conditioners

Simple shampoo

25 g (1 oz/$^1/_3$ cup) (tightly packed)
  soap flakes
2 teaspoons borax
1 litre (35 fl oz/4 cups) very hot
  distilled water
40 drops lavender essential oil
40 drops rosemary essential oil
20 drops basil essential oil (optional)

Dissolve the soap flakes and borax in the hot water. When the mixture has cooled slightly, add the essential oils and stir really well to distribute them through the mixture.

The mixture may go lumpy on standing, so another good stir will be needed before use. The shampoo can be decanted into a squeezebottle or can simply be scooped out of the mixing jar as you need it. The squeezebottle method is best, as this slippery mixture has a habit of sliding off your hands and hair. Follow with the Floral vinegar rinse overleaf.

Castile shampoo

In the following recipe you can use either fresh herbs or essential oils, but remember that a shampoo using fresh herbs won't last as long without preservation as one that uses essential oils.

3 tablespoons finely chopped fresh
  rosemary and lavender
250 ml (9 fl oz/1 cup) distilled water
185 ml (6 fl oz/$^3/_4$ cup) liquid Castile soap
or
20 drops lavender essential oil
10 drops rosemary essential oil
185 ml (6 fl oz/$^3/_4$ cup) liquid Castile soap

For the herbal option, simmer the herbs and water in a covered pan for 30 minutes. Stand overnight if possible. Strain. Pour back into the pan and simmer, covered, until reduced to 60 ml (2 fl oz/$^1/_4$ cup). Strain through coffee filter paper, add to the Castile soap, mix and then bottle. Follow with the Floral vinegar rinse recipe overleaf.

For the essential oil option, add the essential oils to the Castile soap and mix well. Up-end the bottle a couple of times before use. Follow with the Floral vinegar rinse overleaf.

Rosemary is often used in haircare products as it stimulates the circulation.

A vinegar rinse will leave your hair free
of all traces of shampoo. The vinegar
smell will quickly evaporate away.

Anti-dandruff shampoo

Use either the Simple shampoo or the
Castile shampoo on the previous page as
the base for this recipe.

250 ml (9 fl oz/1 cup) shampoo base
20 drops rosemary essential oil
20 drops eucalyptus essential oil
20 drops lemon essential oil

Mix all the ingredients together really
well. Store in a squeezebottle.

To use, first invert the bottle a few
times to mix the oils properly with the
shampoo. Shampoo your hair thoroughly,
paying special attention to the scalp and
massaging with the pads of your fingers,
not the nails. Rinse well. Follow with the
Floral vinegar rinse.

Floral vinegar rinse

40 ml (1^3/$_8$ fl oz) vinegar (white wine or
    apple cider)
20 drops lavender essential oil
20 drops rosemary essential oil
10 drops geranium essential oil (or lemon
    essential oil if your hair is oily)
distilled water

Mix the vinegar and essential oils
together in a 300 ml (10^1/$_2$ fl oz)
spraybottle. Top up with distilled water.
Shake well before use.

To use, rinse your hair after shampooing,
then spray thoroughly with the vinegar
rinse. Don't rinse out.

Rosemary conditioner

All hair types except oily

1 egg, beaten
1 teaspoon vegetable glycerine
2 drops castor oil
2 drops rosemary essential oil
2 drops lavender essential oil
skim milk powder

Beat all the ingredients together, adding
enough milk powder to form a soft paste.

To use, massage into your hair after
shampooing, leave on for a few minutes
and then rinse your hair lightly with
lukewarm water.

Lemon conditioner

Oily hair

2 tablespoons vodka
1 teaspoon runny honey
1 egg white, beaten
2 drops lemon essential oil
1 drop rosemary essential oil
skim milk powder

Beat all the ingredients together with
enough milk powder to form a soft paste.

To use, massage into your hair after
shampooing, leave on for a few minutes
and then rinse your hair lightly with
lukewarm water.

bath time

Herb and oats bath mixture

This mixture features skin-softening oats.

55 g (2 oz/2 cups) dried lavender heads
65 g (2¼ oz/1 cup) dried rosemary leaves
30 g (1 oz/1 cup) dried peppermint leaves
20 g (¾ oz/½ cup) dried comfrey leaves
   (comfrey is a restricted substance in
   some countries)
110 g (3¾ oz/1 cup) rolled oats
10 drops peppermint essential oil

Crumble the dried herbs until the mixture resembles tea leaves. Mix in the oats. Drizzle the essential oil over the herbs and oats, stirring gently. Store the mix in a tightly sealed jar. To use, add to your bath 40-90 g (1½-3¼ oz/¼-½ cup) of the mix tied in a muslin bag.

Floral bath mixture

45 g (1⅝ oz/2 cups) scented dried rose
   petals
60 g (2¼ oz/2 cups) dried rose geranium
   leaves
10 g (¼ oz/½ cup) dried lavender
   flowers and leaves
15 drops ylang-ylang or jasmine
   essential oil

Crumble the herbs until the mix resembles tea leaves. Drizzle the essential oil over the herbs, stirring gently. Store the mix in a tightly sealed jar. To use, add to your bath 40-90 g (1½-3¼ oz/¼-½ cup) of the mix tied in a muslin bag.

Bath herb 'soup'

Fresh herbs in the bath hydrate and soothe the skin but it's not much fun having a bath with twigs and leaves floating around in it. Instead, use the following method.

Put a handful of fresh herbs in a pan. Cover with water, then add 1-2 tablespoons cider vinegar. Cover with a lid and simmer over a low heat for 20 minutes. Strain (you can add the herbs to your compost bucket) and pour the rich herb soup into the bath or sponge it over your body when you have finished showering. The cider vinegar creates the correct acid balance for your skin and also extracts more properties from the herbs than water alone would do.

Golden bubble bath

125 ml (4 fl oz/½ cup) good-quality
   golden-coloured organic shampoo
   (purchase from a health food store)
1 egg, beaten
1 teaspoon runny honey
1 teaspoon vegetable glycerine
20 drops essential oil of your choice

Combine all the ingredients, then bottle. Keep refrigerated and use within 2 weeks. To use the mixture, slowly trickle about 60 ml (2 fl oz/¼ cup) under a fast-running tap to maximize the bubbles.

Keep the bath water temperature to 35°C (95°F). Any hotter will raise your pulse rate, leaving you feeling tired.

citrus refresher bath mixture

This sprightly bath blend relieves that frazzled feeling after a hot day.

60 g (2 1/4 oz/2 cups) dried lemongrass
10 g (1/4 oz/1 cup) dried lemon verbena leaves
25 g (1 oz/ 1/2 cup) dried mint leaves
1 tablespoon dried, ground lemon peel
1 tablespoon dried, ground orange or mandarin peel
10 drops peppermint essential oil

Crumble the dried herbs and citrus peel until the mixture resembles tea leaves in texture.

Drizzle the essential oil over the dried herbs, stirring the mixture gently.

Store in a tightly sealed jar.

To use, add 40-90 g (1 1/2-3 1/4 oz/1/4-1/2 cup) of the mixture tied in a muslin bag to each bath.

Run the hot water over the mixture and then add cold water until the bath is the correct temperature.

bath oils

St Clements bath oil

2 teaspoons lemon essential oil
1 teaspoon orange essential oil
1 drop essential oil of cloves
4 teaspoons tincture of benzoin
125 ml (4 fl oz/$^1/_2$ cup) vodka
50 g (1$^3/_4$ oz) anhydrous lanolin
4 teaspoons runny honey
375 ml (13 fl oz/1$^1/_2$ cups) herb oil base

Dissolve the essential oils and tincture
in the vodka. Heat the lanolin and honey
in a double boiler. Don't overheat. Take
off the heat and cool to 45°C (113°F).
Slowly stir in the vodka and oil blend.
Bottle. Add about 60 ml (2 fl oz/$^1/_4$ cup)
to the bath as the water is running.

Intensive treatment bath oil

125 ml (4 fl oz/$^1/_2$ cup) brandy or vodka
$^1/_2$ teaspoon petitgrain essential oil
$^1/_2$ teaspoon bergamot essential oil
$^1/_2$ teaspoon rosemary essential oil
$^1/_2$ teaspoon lemon essential oil
$^1/_2$ teaspoon patchouli essential oil
2 teaspoons vegetable glycerine
2 teaspoons sweet almond or macadamia oil

Put all ingredients in a bottle. Shake
well, and leave for 3 days to blend. Pour
1 teaspoon only on the surface of the bath
just before you step in. A potent oil!

Lavender in a bath can relieve
dry, itchy skin and relax
tired muscles.

Lavender bath oil

1 egg
2 teaspoons cider vinegar
40 ml (2 tablespoons) vodka
2 teaspoons vegetable glycerine
125 ml (4 fl oz/^{1}/$_2$ cup) light olive or
    macadamia oil
20 drops lavender essential oil
10 drops patchouli essential oil
5 drops rosemary essential oil
1 tablespoon homemade shampoo (page 55)

Beat the egg, cider vinegar, vodka,
glycerine, oil and essential oils together
in a jug. Stir in the shampoo thoroughly,
then bottle. Keep refrigerated and use
within 2 weeks.

To use, add about 60 ml (2 fl oz/^{1}/$_4$ cup)
to the bath as the water is running.

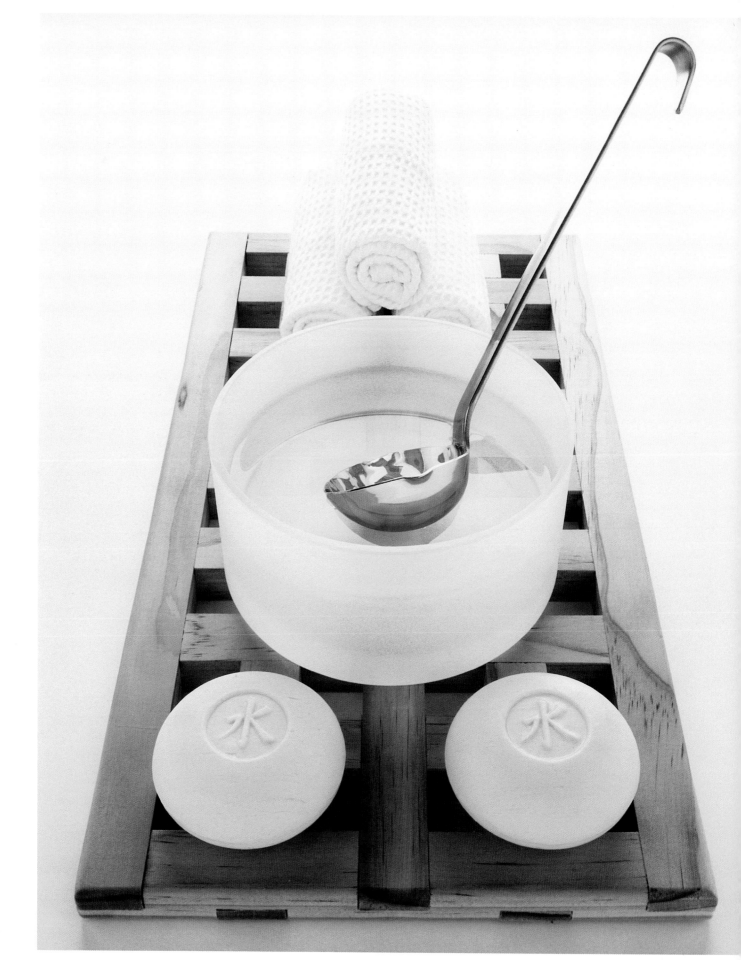

aromatherapy bath blends

The following blends are sufficient for one bath. To avoid floating 'hot spots' of unmixed essential oil, it is best to mix the essential oils with 1 tablespoon of either full-cream milk or sweet almond oil before adding them to the bath.

Skin deodorizing bath blend

4 drops clary sage essential oil
2 drops eucalyptus essential oil
2 drops tea tree essential oil
2 drops peppermint essential oil

Dry skin blend

4 drops chamomile essential oil
4 drops geranium essential oil
2 drops patchouli essential oil

Oily skin blend

5 drops lemon essential oil
3 drops ylang-ylang essential oil
2 drops cypress essential oil

Spotty skin blend

2 drops eucalyptus essential oil
2 drops thyme essential oil
4 drops lavender essential oil
2 drops chamomile essential oil

Aromatic blend

1 drop lavender essential oil
2 drops grapefruit essential oil
2 drops geranium essential oil
2 drops ylang-ylang essential oil
2 drops patchouli essential oil

Relaxing blend

4 drops chamomile essential oil
3 drops lavender essential oil
3 drops ylang-ylang essential oil

Use natural materials in your bathroom, for example cotton bathmats, towels and flannels.

bath salts

Sandalwood salts

100 g ($3^{1}/_{2}$ oz/$^{1}/_{2}$ cup) tartaric acid
125 g ($4^{1}/_{2}$ oz/$^{1}/_{2}$ cup) bicarbonate of soda
30 g (1 oz/$^{1}/_{4}$ cup) cornflour
20 drops geranium essential oil
2 drops grapefruit essential oil
4 drops sandalwood essential oil

Mix the dry ingredients together. Add the essential oils drop by drop, stirring all the time to prevent caking. Put the mixture in a jar and shake daily for a few days to blend. To use the salts, sprinkle about 30–60 g (1–$2^{1}/_{4}$ oz/$^{1}/_{4}$–$^{1}/_{2}$ cup) of the mixture in the bath as it is running.

Scent of the forest salts

125 g ($4^{1}/_{2}$ oz/$^{1}/_{2}$ cup) bicarbonate of soda
140 g (5 oz/$^{1}/_{2}$ cup) Epsom salts
100 g ($3^{1}/_{2}$ oz/$^{1}/_{2}$ cup) tartaric acid
140 g (5 oz/1 cup) Irish moss, ground to
    a powder
850 g (1 lb 14 oz/3 cups) fine sea salt
4 drops pine essential oil
20 drops eucalyptus essential oil

Mix the dry ingredients together. Add the essential oils drop by drop, stirring to prevent caking. Put the mixture in a jar and shake daily for a few days to blend. To use the salts, sprinkle 30–60 g (1–$2^{1}/_{4}$ oz/$^{1}/_{4}$–$^{1}/_{2}$ cup) of the mixture in the bath as it is running.

Geranium essential oil
is good for poor circulation,
and is also said to help
balance mood swings.

exfoliation treatments

Dry-brush massaging

A dry-brush massage is a terrific treatment for your skin. It stimulates the circulation and gets rid of the dead cells that can make skin look dull; skin feels very 'alive' after brushing. It's possible to buy inexpensive, long-handled brushes at pharmacies and variety stores. Use the brush before your bath or shower and brush in a circular movement all over your body, paying special attention to areas where there are glands, such as the armpits, groin and just below the collarbone. Be gentle on breasts and around the genital area.

Body scrub

This scrub will thoroughly cleanse the skin. You can use either salt or sugar — try both to see which you prefer. The salt and sugar exfoliate the skin and the oil moisturizes, so you come out of the shower with skin that feels like that of a new baby.

4-5 teaspoons sugar or salt
1 teaspoon light olive oil
1 drop essential oil of your choice

Mix the ingredients together very well.

To use, wet your whole body in the shower, then turn the water off. Apply the sugar or salt scrub and massage it into your skin. Don't use it on your face until you are sure that it's not too abrasive. Rinse well.

Normal skin sheds old cells and renews them every 28 days, but this process slows as you age. Exfoliating your skin will help restore its healthy glow.

natural home

environmentally considerate choices, for your home and how you live in it

As with any investment, we have an instinctive desire to protect our homes. We clean them, decorate them, renovate them. Yet so many of us heedlessly pollute our personal spaces on a daily basis without the slightest awareness of the potential harm we are doing to ourselves. In the quest for a shiny, fragrant home we regularly apply a cocktail of chemicals and poisons to our household surfaces as well as releasing them into the air that we breathe.

While the message that stronger can mean harsher and more poisonous is definitely out there, there seems to be a general reluctance to acknowledge it fully. Somehow we assume that natural means ineffectual and gentle means weak. But there is a way to clean and maintain your home without turning it into a toxic battleground, a way that works for you and the environment. There is a whole arsenal of natural ingredients and techniques at your disposal that will enable you to keep your home squeaky clean and poison free. With this chapter as your guide you'll discover that there is no need for a trade-off between your standards and your peace of mind.

the green house

An environmentally friendly house is an energy-efficient and comfortable one — warm in winter, cool in summer, light-filled key rooms and pleasant to be in because it is well ventilated and smells fresh. The green house makes gentler demands on the environment, and the pocket.

Ideally, the green house is positioned to take maximum advantage of the sun's light and warmth year-round. It relies less on non-renewable energy sources (such as coal) by making the best use of low-impact, renewable energy sources, such as sun and wind. It conserves water and reuses it when possible. The green house is a low-polluting one, not only in terms of the energy sources it uses, but also in the products that are used within its walls. The green householder chooses low-toxic, natural cleaning products in preference to harsh and highly polluting chemicals.

Of course, it's often not possible to build a house from scratch, and sometimes you may just have to do the best you can with what you've got. Here are a few energy-efficient principles to get you started.

- Use the sun and wind where possible for energy. This can be as simple as pegging out the washing rather than using your dryer.

- Think of the local environment: plant native trees and flowers, compost organic wastes, garden organically and use natural pest control. Use low-flush or waterless toilets, and collect, store and use rainwater.

- Use green materials and products. The green ideal is to use non-toxic, non-polluting products from sustainable and renewable sources, which are biodegradable or easily reused and recycled.

- Paint the exterior of a house with light-coloured paint to help reflect unwanted radiant heat.

- Create a healthy indoor climate by allowing the house to breathe. Use natural materials and processes to regulate the temperature, humidity and airflow.

- Allow sunlight and daylight to penetrate your home, and rely less on artificial lighting.

green house logistics

Once you make a commitment to 'going green', you'll be surprised at how quickly even small savings turn into big ones — both for you and for the environment.

* Install a solar or energy-efficient hot water heater. Since water heating accounts for up to 50 per cent of a home's energy use, installing one of these heaters saves in energy bills as well as pollution.

* Choose energy-efficient appliances when buying new ones. Many countries have star ratings to make the choice easier. Top-rated refrigerators, freezers, washing machines, dryers, dishwashers and air-conditioners are much less polluting and also cheaper to run.

* Install a water-efficient showerhead. These usually pay for themselves in the first year of use. The shower is the largest user of household hot water and accounts for around 20 per cent of the greenhouse pollution in the average home. These showerheads use less water, reducing both pollution and heating costs.

* Consider ways of adding thermal mass to your home if you live in a climate with cool winters. Thermal mass describes heavy building materials such as brick, stone or thick ceramic tiles that are slow to heat and slow to cool. In the winter they warm up in the day and continue to radiate heat in the evening, while in summer they protect against excessive heat, especially when shaded. New brick, tile or concrete flooring is an obvious way of adapting an existing home to include greater thermal mass.

* Use appliances only when you really need them. For instance, use a broom in the garden, not a motorized leaf blower.

- Use appliances efficiently and maintain them well so they work optimally.

- Don't leave the refrigerator door open unnecessarily. For every minute it is open, it takes 3 minutes to cool down again.

- Switch off lights when they are not in use.

- Replace your most frequently used light bulbs with compact fluorescents. Each bulb uses around a quarter of the electricity needed for a standard bulb and will prevent the emission of half a tonne (half a ton) of greenhouse pollution over the life of the bulb. If you fitted all your lamps and lights with energy-efficient bulbs, you could reduce your lighting costs by 80 per cent.

- Turn off microwaves, TVs and sound systems at the power point, as they use power even when not operating.

- Continue to wash clothes in cold water if you're satisfied with the result. Consider pre-soaking heavily soiled garments first.

- Wear adequate clothing when the weather cools and don't attempt to heat your entire house to the same even temperature. Keep one or two rooms cosy by shutting the doors to the rest of the house.

Consider fitting a flow restriction disc to your current shower rose.

hot water wise

Heating water burns up money as well as producing pollution. Try these simple ideas for reducing your hot water usage.

- Take quicker showers. Cutting your shower time from 10 minutes to 5 could save as much as 27 000 litres (5939 gallons) of water a year.

- Fix dripping taps. Forty-five drops a minute is 10 baths of water a year.

- Don't rinse dishes under a running hot tap. Use a sink or bowl full of water instead.

- Avoid turning the hot tap on for small quantities of water. This leaves the pipes full of hot water, which cools and is wasted.

- Turn off the hot water system if you'll be away from home for more than 2 or 3 days.

- Insulate your storage tank. Between 15 and 20 per cent of the cost of running an electric hot water service is due to heat losses from the storage tank. Reduce losses by wrapping the tank in foil-backed insulating blanket held in place with ducting tape. (Do not add extra insulation to gas systems as they may overheat.)

- Install an on-demand hot water system if you have a choice, as they are the most efficient. You are not paying to heat stored water.

- Locate new hot water systems near where you use hot water to cut the amount of hot water and energy lost through cooling in pipes.

To measure how much water your average shower uses, hold a bucket under the shower for 10 seconds. Measure this volume and multiply it by 6 to give you the number of litres (gallons) per minute.

The other key component
of the green cleaning
kit is elbow grease. But
you won't need too much
of it if your approach
is a small amount of
cleaning effort on a
regular basis.

green cleaning

There is no easier way to start making your home more environmentally friendly than by considering the substances that you use to maintain it. On the following pages are some homemade cleaning options, including some stronger mixtures for those who are making a progressive transition to natural housekeeping.

your basic green ingredients

Bicarbonate of soda

Also known as sodium bicarbonate and baking soda, this gentle, moderately alkaline, non-toxic abrasive cuts through grease and oil because it reacts with the fatty acids to form mild detergents. Use it to clean, deodorize and buff. You can even use it to clean your teeth!

Lemon juice

This natural bleach can be used for many purposes, such as removing stains, deodorizing, and inhibiting mould.

Salt

A natural gentle abrasive and disinfectant, salt is useful for clearing drains and scouring kitchen utensils.

White vinegar

This type of vinegar has a key role in alternative home cleaning because of its acidic properties, although it is not a general cleaner. This moderately strong acid can remove bathroom scum and hard water deposits as well as discolouration from metals such as aluminium, brass and copper. It can also remove rust stains and rust on iron. White vinegar makes an ideal cleaner for the bathroom.

home-brewed cleaners

All-purpose cleaner

2 heaped tablespoons bicarbonate of soda
1 tablespoon white vinegar

Mix the bicarbonate of soda and white vinegar together and store the cleaner in an airtight container. To use, wipe surfaces with a soft cloth dipped in the solution. Rinse with clean water.

Mild all-purpose cleaner

4 tablespoons bicarbonate of soda
1140 ml (2^1/$_2$ pt) warm water

Mix the ingredients together. To use, wipe on surfaces with a soft cloth and rinse with clean water.

Strong all-purpose cleaner

Do not use on fibreglass or aluminium.

110 g (3^3/$_4$ oz/^{1}/$_2$ cup) washing soda crystals
4.5 litres (9^1/$_2$ pt) warm water

Mix the ingredients together. To use, wipe on surfaces with a soft cloth and rinse with clean water.

Mild abrasive cleaner

Use this cleaner on plastic and on painted walls.

a few drops of water
bicarbonate of soda

Add a few drops of water to bicarbonate of soda to form a paste. To use, apply it with a stiff-bristled brush on hard surfaces, and with an old toothbrush between tiles.

Scouring cleaner 1

1 teaspoon borax
2 tablespoons white vinegar
500 ml (17 fl oz/2 cups) hot water

Combine the ingredients and pour the mixture into a spraybottle.

Scouring cleaner 2

125 g (4^1/$_2$ oz/^{1}/$_2$ cup) bicarbonate of soda
3 tablespoons sodium perborate

Mix the bicarbonate of soda with the sodium perborate.

To use, rub the mixture onto areas that need whitening with a wet sponge. Leave for 10 to 15 minutes before rinsing.

A cotton mop will last longer than a synthetic refillable sponge.

Disinfectant 1

1 teaspoon borax
2 tablespoons distilled white vinegar
60 ml (2 fl oz/1/4 cup) liquid soap
500 ml (17 fl oz/2 cups) hot water

Mix the ingredients together. To use, store in a spraybottle and spray it on, then rinse and allow to dry.

Disinfectant 2

2 teaspoons borax
4 tablespoons white vinegar
750 ml (26 fl oz/3 cups) hot water

Mix the ingredients together. Pour the mixture into a spraybottle. For greater cleaning power, add 1/4 teaspoon liquid soap.

Disinfecting floor cleaner

100 g (3 1/2 oz/1/2 cup) borax
3 litres (6 1/2 pt) hot water

Mix the borax and hot water together. Use the cleaner with a cloth or mop.

Basic dishwashing liquid

50 g (1 3/4 oz) pure soap
5 litres (10 1/2 pt) cold water
110 g (3 3/4 oz/1/2 cup) washing soda crystals
1 1/2 teaspoons eucalyptus oil or tea tree oil
125 ml (4 fl oz/1/2 cup) white vinegar
a few drops lemon or lavender pure
   essential oil for fragrance
4.5 litres (9 1/2 pt) hot water

Grate the soap into a large saucepan and cover with 1 litre (35 fl oz/4 cups) of the cold water. Bring to the boil, add the washing soda and stir until it is dissolved. Stir in the eucalyptus oil, vinegar and essential oil. Pour into a bucket, add the hot water, then stir in the remaining cold water. When cool, transfer to smaller containers and label.

To use, add 1 teaspoon to 5 litres (10 1/2 pt) water when washing up, or use 250 ml (9 fl oz/1 cup) per load in a dishwasher. (Note: It does not remove coffee and tea stains.)

Strong floor cleaner

Wear protective gloves when using this cleaner. Do not use it on waxed floors.

50 g (1 3/4 oz/1/4 cup) washing soda crystals
1 tablespoon liquid soap
125 ml (4 fl oz/1/2 cup) white vinegar
3 litres (6 1/2 pt) hot water

Mix the ingredients together. Use the cleaner with a cloth or mop.

Pure soap is 100 per cent biodegradable and does not pollute the environment.

kitchen

Here are a few tried and true tips, both old and new, to keep everything in your kitchen sparkling clean.

Lemon fresh

To wash and deodorize the refrigerator, garbage bin or kitchen compost container, use a solution of 1 teaspoon lemon juice to 1 litre (35 fl oz/4 cups) water.

Tarnished silver cutlery

A quick fix is to place a piece of aluminium foil in a plastic bucket and sprinkle over 3 tablespoons bicarbonate of soda. Lay the silver on top. Cover with hot water. Leave until the bubbles stop, rinse and polish with a dry soft cloth.

For sparkling crystal

Dip crystal in a solution of 1 part vinegar to 3 parts water. Polish with a dry, lint-free cloth.

Washing glass

Wash glass water bottles and stained flower vases with 1 tablespoon vinegar and 1 tablespoon salt in warm water. Allow to soak for several hours and shake occasionally.

Delicate china

When hand washing the heirloom china, place a towel at the bottom of the washing up bowl to help prevent chips and breakages.

Wash silver cutlery as soon as possible. Some food stains are harder to remove using gentle green methods if you leave them too long.

Removing wax from candlesticks

Hot method: Re-melt wax with a hair dryer on the hot setting, and wipe the wax as it softens.

Cold method: Place candlesticks in the freezer for about an hour. This makes it easy to peel the wax off.

Burnt pans

To clean a badly burnt saucepan, pour in a little olive oil, heat gently and leave to stand for 1 or 2 hours. Pour off the oil into a container, ready to use for the next burnt pan, and wash the pan as usual.

To clean burnt food from a pie dish, dip it in very hot water, then quickly turn it upside down onto a flat surface. This traps steam, which loosens the residue.

To restore an enamel baking dish that seems burnt beyond use, soak it in a mixture of water and strong soap powder. After a couple of hours, pour off the water and rub the dish with a soft cloth.

Greasy grill pans

When you've scraped off as much grease as you can with a spatula or newspaper, sprinkle with washing soda crystals and pour on boiling water. Leave it to soak for at least 10 minutes, then clean up. The grease and debris will just lift off.

Stained enamel

To whiten enamel-lined pans, finely crush eggshells and rub with a cloth dipped in salt to remove stains.

You can remove some stains from glassware by gently rubbing the item with half a lemon, seeds removed.

oven cleaners

When you do have to clean the oven, try one of these alternatives to harsh proprietary cleaners.

Natural oven cleaner 1

Wet the surface and sprinkle it with bicarbonate of soda. Rub with fine steel wool. Wipe off the residue with a damp cloth. Repeat if necessary. Rinse well and dry.

Natural oven cleaner 2

To clean very dirty shelves, soak them in a mixture of 1 part washing soda to 4 parts hot water. If the shelves are too large to submerge fully in the sink, turn them around every 20 minutes, or use the bath or laundry sink.

Strong oven cleaner

Preheat the oven to warm, or proceed immediately after cooking in the oven. Place 125 ml (4 fl oz/$1/2$ cup) cloudy ammonia inside, shut the door and switch off the oven. Leave overnight if possible, or for at least a few hours. Wipe thoroughly with hot water and detergent.

To avoid the need for harsh caustic oven cleaners, don't let the oven get too dirty. Wipe surfaces with a hot damp cloth after each use.

Natural fragrances such as vanilla
smell better than artificial air
fresheners and contain no high-
irritant chemicals.

appliances

Bicarbonate of soda mixed with water is one mild, non-toxic and environmentally friendly option for cleaning kitchen appliances, but here are some more specific tips.

Microwaves

For stubborn stains, place a bowl of hot water in the microwave and switch on to high for about 5 minutes. Leave the bowl to stand for a few minutes and then remove. Wipe inside the microwave with a soft cloth.

Dishwasher

Clean the filters and seals on your dishwasher regularly.

To restore a dull interior, run the machine empty on a short cycle with 500 ml (17 fl oz/2 cups) white vinegar in the detergent receptacle.

Refrigerator

Use a solution of 1 part bicarbonate of soda to 7 parts water to wipe down the inside of the fridge. Wash any removable parts in hot water and detergent.

To leave the fridge smelling fresh, wipe over the inside with a damp cloth and a few drops of vanilla essence.

To prevent mould forming on the fridge door seals, wipe over them with white vinegar. To prevent mould growing when your fridge and freezer are empty and switched off for more than a few days, wipe the inside with vinegar and leave the doors propped open.

If you can reach them, vacuum the coils behind the fridge using your vacuum cleaner's brush attachment.

To absorb odours, place a small open bowl of bicarbonate of soda on one of the shelves. Change it regularly.

A natural way to de-scale your kettle

Cover the element with vinegar, then top up with water. Bring to the boil and then leave overnight, preferably for about 12 hours. Pour the liquid away.

natural drain cleaners

When the water won't drain and you are faced
with a sink full of water, try one of these
natural drain cleaners, then use a plunger.
Place the plunger tightly over the drain
hole, push down and then pull up rapidly,
keeping the plunger over the hole. If your
seal is tight, the air and water inside the
pipe is forced back and forth, with any luck
sloshing and sucking the blockage away. If
you are unsuccessful, try a different drain
cleaner and leave overnight.

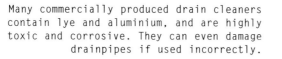

Many commercially produced drain cleaners
contain lye and aluminium, and are highly
toxic and corrosive. They can even damage
drainpipes if used incorrectly.

Sodium bicarbonate and vinegar

125-250 g (4$^1/_2$-9 oz/$^1/_2$-1 cup) bicarbonate
    of soda
250 ml (9 fl oz/1 cup) white vinegar

Pour the sodium bicarbonate down the
drain, then slowly pour in the white
vinegar. The sizzling sound is the
reaction between the two. Follow this
with a flushing of water and repeat the
whole process if necessary.

Washing soda and boiling water

450 g (1 lb/2 cups) washing soda crystals
a kettle of boiling water

Pile the washing soda at the mouth of
the drain, then slowly pour on the
boiling water.

laundry

alternative washing products

Basic household soap cleaner

This cleaner can be used as a laundry detergent. It's suitable for machine washing or hand washing, for front loaders and top loaders. Use approximately 500 ml (17 fl oz/2 cups) per load. It can also be used as a prewash. Soak heavily soiled items in a solution before washing.

50 g (1^3/$_4$ oz) pure soap, grated
5 litres (10^1/$_2$ pt) cold water
110 g (3^3/$_4$ oz/^{1}/$_2$ cup) washing soda crystals
1^1/$_2$ teaspoons eucalyptus oil or
    tea-tree oil
125 ml (4 fl oz/^{1}/$_2$ cup) white vinegar
a few drops lemon or lavender essential
    oil for fragrance
4.5 litres (9^1/$_2$ pt) hot water

Grate the soap into a large saucepan and cover with 1 litre (35 fl oz/4 cups) of the cold water. Bring to the boil, add the washing soda and stir until completely dissolved. Stir in the eucalyptus oil, white vinegar and essential oil. Pour the mixture into a bucket, add the hot water, then stir in the remaining cold water. When cool, transfer to smaller containers and label.

All-purpose laundry stain remover 1

Use this stain remover for soaking soiled clothing. Alternatively, apply it with a sponge and leave to dry. Then wash the item as usual.

50 g (1^3/$_4$ oz/^{1}/$_4$ cup) borax
500 ml (17 fl oz/2 cups) cold water

Mix the borax and water together.

All-purpose laundry stain remover 2

3 tablespoons eucalyptus oil
250 ml (9 fl oz/1 cup) methylated spirits
250 ml (9 fl oz/1 cup) boiling water

Mix the ingredients together in an enamel or stainless steel bowl. Stand the bowl in a bigger saucepan of hot water over medium heat, stirring the mixture until it turns transparent. Pour into moulds such as empty milk or juice cartons. When the mixture has set, cut it into bars and leave for 4 weeks to harden.

The average person generates over a tonne (ton) of dirty clothes every year.

recycling wool

Recycling wool from old, misshapen or unfashionable jumpers is a sadly forgotten craft. Recycled wool is also called 'shoddy', now a derogatory term that seems inappropriate since recycling has lost its negative image to a large degree.

As you unravel the wool, wind it round the back of a chair or a large piece of cardboard.

To revive the wool, steam it by holding it over hot, steaming water with two wooden spoons until it is evenly damp. Alternatively, hang it in the shower recess while you have a hot shower.

Leave the wool until it is dry.

Wind the wool into balls. Dye it or use it as is.

Recycled wool is ideal for children's knitting experiments as well as more ambitious projects.

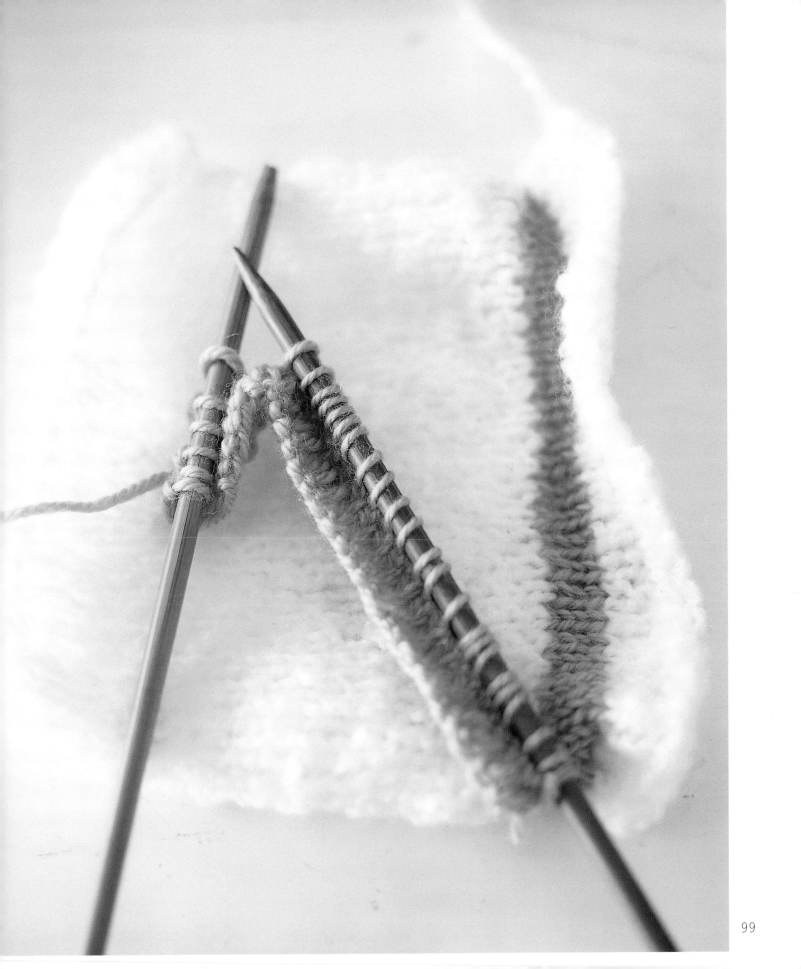

Eucalyptus is often used for its
scent, but it is also a proven
stain remover.

old-fashioned laundry aids

Fragrance

Add fragrance to a washing load by adding a few drops of lavender essential oil to a face cloth and dropping it into the machine. You could also try ylang-ylang, neroli, pine or eucalyptus oil.

Softness

Clothes will be soft to the touch if you soak them overnight in a solution of 1 part vinegar to 3 parts water. Rinse well before washing.

Lint prevention

A cup (about 250 ml/9 fl oz) of white vinegar in the final rinse water in a load of washing will help prevent lint from forming.

Soap

As soap is 100 per cent biodegradable and does not pollute the environment, environmental groups recommend it for washing clothes in preference to washing detergents. Available in supermarkets, soap flakes are ideal for the gentle washing of clothes — for instance, silks and wools — that may be damaged by strong detergents. Laundry soap is also available in blocks and may contain additives such as borax, bleach, sodium phosphates and perfumes. Unlike synthetic detergent, soap does not dissolve well in cold water.

Washing soda

Otherwise known as sodium carbonate, washing soda is a mild alkali available in a crystal or powder form. It is a good water softener and stain remover and is used as an additive in commercial laundry products. Use it with soap for laundry, as a tarnish remover on silver and to help unblock drains.

Ironing

The most basic ironing aid you can use is a damp cloth, preferably white cotton as this won't result in colour run or leave behind fluff or lint. The cloth protects the fabric as you iron.

dry-cleaning

Salt, bran and clay are just three cleaning agents used in the past to care for garments that could not be laundered. Some of the methods require a little space and could prove quite messy. Choose an outdoors location or an easy to clean room such as a tiled bathroom, and protect your hair with a scarf.

General purpose dry-cleaner

Water mixed with fuller's earth (an absorbent clay with a gentle cleaning action) will remove grease and oil from materials such as felt. Mix the fuller's earth with water to form a paste. Spread it on the article. Let it dry, then remove it with a stiff brush.

Light and delicate items

Place the item on a clean towel, then rub all over it with French chalk, giving extra attention to marked areas. Roll the item up in the towel and leave it for 3 to 4 days. Unwrap and brush lightly to remove the chalk.

Leather jacket

Give a leather jacket a facelift by covering it with a paste made from pipe clay (fine white pure clay) and water. Use just enough water to make the clay spreadable. Rub it in one direction only, from the bottom to the top. Allow the paste to dry, then shake the garment until all the clay has dropped off.

Woollens

Salt applied with a linen pad is suitable for cleaning woollen dresses and skirts. Lay the garment on a table and sprinkle a thin layer of salt over it. Spread it evenly with your fingers and then, with a linen pad (a piece of linen folded several times), rub the salt into the cloth in long sweeping movements towards the hem. (A circular motion may roughen the surface.) Place the garment on a clothes hanger and brush it vigorously with a stiff clothes brush. Cuffs, collars and hems may need a repeat treatment.

For white or light-coloured wool, you could substitute ground rice for the salt, but you will need to leave the rice powder on for several hours.

Correct clothes care will help you avoid costly dry-cleaning bills. Brush clothes to remove lint and surface dirt and occasionally air them outside.

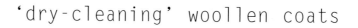

'dry-cleaning' woollen coats

You don't have to head straight to the dry-cleaners if you have a woollen coat or jacket that can't be washed using soap and water. The following method is not recommended for pale-coloured items, however.

Mix together 250 g (9 oz) of flour and 150 g (5$\frac{1}{2}$ oz) bran (the flaky outer husks of the wheat grain).

Heat the mixture in the oven, turning frequently to prevent it from browning.

When the mixture is as hot as possible without burning, spread it over the fabric of the garment.

Fold the hot garment in a towel and put it aside.

Two or three days later, unwrap the garment, shake it energetically and brush with a clothes brush to remove all trace of the mixture.

bathroom

natural fresheners

To freshen the air in the bathroom and remove odours, make sure it has adequate ventilation. If necessary, open the window and try one of the following non-toxic air fresheners.

- Put 60 ml (2 fl oz/$^1/_4$ cup) white vinegar in an open bowl positioned on a high shelf.

- Do the same with a bowl of clay-type cat litter.

- Make a spray air freshener by combining 1 teaspoon bicarbonate of soda and 1 teaspoon lemon juice in 500 ml (17 fl oz/2 cups) hot water and decanting into a pump spraybottle.

- Use scented candles to help banish bathroom smells.

- Fragrant lemon essential oil kills germs, so add a couple of drops to the final rinse water when cleaning the bathroom.

Antiseptic air freshener

$^1/_4$-$^1/_2$ teaspoon any antiseptic essential oil (choose from thyme, bergamot, juniper, clove, lavender, peppermint, rosemary or eucalyptus)
1 teaspoon methylated spirits
500 ml (17 fl oz/2 cups) distilled water

Dissolve the essential oil in the methylated spirits and then blend this with the distilled water. Pour into a pump spraybottle. Use your customized air freshener on the fine mist setting.

Natural toilet cleaner

Try this gentler alternative to commercial toilet products. Add a few drops of pine essential oil to this recipe for additional disinfectant power.

225 g (8 oz/1 cup) borax
60 ml (2 fl oz/$^1/_4$ cup) white vinegar or
    lemon juice

Mix the ingredients together. To use, pour into the toilet bowl. Leave at least a few hours — overnight if possible — then scrub the bowl with a toilet brush.

Bathrooms should be spaces for relaxation and retreat, as well as for cleansing.

Natural disinfectant

20-30 drops of any essential oil with
    disinfectant properties (choose from
    cinnamon, clove, pine, tea tree, thyme,
    bergamot, peppermint, rosemary, juniper
    or sandalwood)
1 teaspoon methylated spirits
1 litre (35 fl oz/4 cups) distilled water

Dissolve the essential oil in the
methylated spirits. Mix with the distilled
water and store in an airtight plastic or
glass bottle.

Soap gel

Recycle your soap scraps into liquid soap
gel that you can use in a recycled pump
pack. It's great for washing your hands
and is useful whenever you need to use
pure soap.

a handful of soap scraps
water
a few drops essential oil of your
    choice (optional)

Place the soap scraps in a saucepan, cover
them with water and leave for 24 hours,
stirring occasionally. Bring to the boil
over moderate heat, stirring continuously,
then lower the heat to a simmer. To
completely dissolve the soap, whisk or
mash it. Remove from the heat and allow
the soap to cool. Mix through a few drops
of essential oil, if desired. Pour into a
pump dispenser for hand washing.

Hair brushes

To clean brushes the old-fashioned way, dissolve a walnut-sized
piece of washing soda in hot water in a basin. Comb any hair out
of the brushes, then dip the brushes, bristles downwards, into the
water, keeping the backs and handles out of the water as much as
possible. Repeat until the bristles seem clean. Rinse in a little
cold water, shake well and wipe the handles and backs, but not the
bristles, with a towel. Place in the sun or near a heater to dry.

Don't use soap on brush bristles,
or wipe them, as this will
soften the fibres.

living areas

caring for your floors

Removing stains from linoleum and vinyl floors

Quick action is always best: wipe, mop and scoop spills as soon as possible. Avoid using abrasive cleaners as they can scratch or dull the surface. If you are left with stains, try the following methods:

*   Wipe the stain with an all-purpose cleaner (choose one of the homemade recipes from the introduction to this chapter).

*   Use lemon and salt for rust stains. Cut a lemon in half, sprinkle with plenty of salt and rub onto the stain. Using a rag or sponge, rinse with water.

*   For thick grease or tar, try mineral spirits, but use it with caution and test it first on a tiny portion (mineral spirits can take the shine off). Another way to remove tar is to cool it with an ice cube and then pry it off with a spatula once it becomes brittle.

*   Organic stains such as blood, grass and pet accidents can be treated with lemon juice, or lemon and salt.

Basic stain-removing foam for carpets and upholstery

50 g (1^3/$_4$ oz) pure soap
5 litres (10½ pt) cold water
110 g (3^3/$_4$ oz/^{1}/$_2$ cup) washing soda
   crystals
1½ teaspoons eucalyptus oil or
   tea tree oil
125 ml (4 fl oz/^{1}/$_2$ cup) white vinegar
a few drops lemon or lavender essential
   oil for fragrance
4.5 litres (9^1/$_2$ pt) hot water

Grate the soap into a large saucepan and cover with 1 litre (35 fl oz/4 cups) of the cold water. Bring to the boil, add the washing soda and stir until completely dissolved. Stir in the eucalyptus oil, vinegar and essential oil. Pour into a bucket, add the hot water, then stir in the remaining cold water. Leave to cool.

To use, beat the cool mixture to a light foam and spoon over the stain. Leave it for 10 minutes, then wipe with a sponge dipped in white vinegar to remove the alkalinity left by the cleaning solution. Rinse with plain warm water (by either spraying the liquid onto the carpet or by patting it on with a clean white cloth or paper towel) and blot thoroughly.

When choosing timber flooring, refer to the international system of certification produced by the Forest Stewardship Council, which will help you to select sustainable timbers.

cleaning timber surfaces

Basic timber floor cleaner

Here's a good, safe cleaner for damp mopping your timber floor.

60 ml (2 fl oz/$1/4$ cup) liquid soap
125 ml (4 fl oz/$1/2$ cup) to 250 ml
   (9 fl oz/1 cup) white vinegar or
   lemon juice
9 litres ($9^1/2$ qt) warm water

Mix all of the ingredients together in a bucket.

Apply with a mop or a sponge mop. Rinse with clean water. For the definitive floor wash, change the washing water as soon as it is dirty (otherwise all you are achieving is spreading dirty water over the floor), and rinse with fresh water and a well cleaned (or a second) mop. Change the rinsing water if it becomes dirty.

Homemade furniture polish

55 g (2 oz) beeswax
140 ml ($4^3/4$ fl oz) turpentine
2 tablespoons linseed oil
2 tablespoons cedar oil

Grate the beeswax into a heatproof bowl. Pour in the other ingredients. Stir. Place the bowl over a saucepan of simmering water until the ingredients have all melted. Allow to cool before using.

To use, rub sparingly over wooden furniture with a soft cloth.

Beeswax is suitable for all interior wood, lending the timber lustre and golden tones. Pure beeswax conditions and protects wooden surfaces.

wall to wall cleaning

Painted surfaces

To wash painted walls, doors and ceilings, use a mild detergent solution. If you need something stronger, use a solution made with 100 g (3$\frac{1}{2}$ oz) washing soda crystals to 4 litres (8$\frac{1}{2}$ pt) water.

Prepare the room first by laying drop sheets on the floor and over any furniture that may receive an unwelcome shower. Some paints take to a wash better than others. Gloss, for instance, is usually fine but an old-fashioned paint such as whitewash may get washed off itself if it is too harshly treated. If you are in doubt about the washability of your paint surface, test wash an area that is not on public display.

Dry clean the wall with a duster or vacuum cleaner before wetting it. When washing, start from the bottom and work up, rubbing the wall surface briskly with your cleaning solution, rinsing off and drying with a soft towel. Catch drips fastidiously as you go, as dirty drips can be even harder to clean up than the original dirt source. Fortunately, they don't stain a damp wall that's already been cleaned.

To rinse, wipe drips with a clean damp sponge. You will need two buckets — one for your cleaning solution and one for rinsing water. Depending on the area of the wall, you may need to change the water frequently.

Wallpaper

Non-washable wallpaper requires gentler treatment. Dust regularly with a soft brush or cloth. Spot clean by dabbing with powdered borax and brushing out. Another homemade remedy is to rub the wallpaper gently with a piece of bread rolled into a ball, or with a soft rubber eraser.

cleaning windows

There is something very satisfying about seeing sun stream through a gleaming pane of glass. Like any job, you need the best tools to do it properly.

You'll need the following equipment:

- steady ladder to reach higher windows (call a window cleaner if you're unsteady on your feet or suffer from vertigo)

- bucket of cleaning solution (you can use a homemade cleaner from this chapter)

- sponge

- squeegee — a handle with a thin strip of sponge on one side and a rubber strip on the other

- clean cloth

- newspaper

The best sort of day to wash windows is a cloudy one, as sun on the windows causes the glass to dry too quickly and unevenly, resulting in streaks.

If you plan to clean or wash the window frames too, do these before you wash the glass. Wipe or dust the frames first and if that's not enough, follow with a wash and wipe dry.

Starting at the top of the window, wash the window with a cloth or sponge.

Squeegee the surface or dry it with a clean, dry, lint-free cloth.

For really dry, sparkly windows, polish with a few sheets of newspaper.

To prevent windows from fogging, rub over a little glycerine after cleaning them.

Window tips

For a quick clean, when the glass is not too dirty, try wiping windows first with wet newspaper, then with dry. The ink on the newspaper gives them a polish.

Never try to clean a window with a dry cloth as the dirt could scratch the glass.

Never use soap on windows as it leaves smears that are very difficult to remove.

Don't hang newly dry-cleaned
items in your bedroom, as they
will emit harmful vapours.

bedroom

storing clothes

Before putting clothes away for long-term storage, or to store until next season, follow this checklist.

- Do wash or dry-clean clothes. Dirt is more likely to attract pests such as insects, and encourage mildew.

- Do air items, especially ones you've steam ironed or damped down before ironing.

- Do use muslin or canvas storage bags, or clean white or undyed sheets.

- Do place items on wire racks, not shelves, so air circulates properly.

- Don't starch items to be stored, as starch attracts meal-seeking silverfish, which do not discriminate between the starch and the clothing fibre.

- Don't store clothes in dry-cleaning plastic or other garment bags that do not breathe, as moisture may be trapped inside. In addition, dry-cleaning-bag plastic may cause yellow streaks over time.

- Don't put clothes or shoes away while still damp.

- When packing a suitcase or putting away valuable clothes, placing tissue paper (acid-free for storage) between fabric the layers of the garments themselves will help reduce creasing and wrinkling.

moths

The webbing clothes moth (*Tineola bisselliella*) is attracted to wool and furs, and its larvae eat the fibres, leaving holes. Stored, unused clothing is most vulnerable. Commercial moth treatments contain naphthalene and organochlorines.

If you suspect moths have invaded an item, wrap it in a damp towel, and put it in a low oven to steam out the grubs, or place underneath a damp towel and press with a hot iron. To defend drawers and cupboards from moths, keep them free from dust by regularly airing and vacuuming them, and wipe over with a repellent such as eucalyptus oil. Cottonwool buds dipped in essential oils deter moths: try lavender, lemongrass or rosemary and put a couple in each drawer between clothing. Or dot a few drops of essential oil on sheets of blotting paper, fix the scent with orris root powder and use as drawer liners.

Old woollen socks ready for
retirement can be recycled and
used for applying shoe polish.

top shoe tips

- Shoes will last longer if given a day's rest after wearing to allow moisture to evaporate, so have at least two pairs for each season.

- Keep your shoes on shoe trees when not in use so that they keep their shape and don't suffer cracking.

- Caring for the leather helps prevent cracks developing.

- Use a hard brush to remove mud; never scrape shoes with a knife.

- Repair damaged shoes as soon as they need it. Delaying repair is a false economy.

- To deodorize shoes, sprinkle bicarbonate of soda inside them. Leave for a day or two, then shake out and air.

- To waterproof footwear the old-fashioned way, melt equal quantities of beeswax and mutton suet. Rub over the top and soles while still liquid and wipe off the excess.

- To remove mildew on leather, rub vigorously with petroleum jelly.

- To soften leather, rub with lemon juice or castor oil. Olive oil helps to prevent the leather from cracking and drying.

- To help break in new shoes, carefully pour a small amount of methylated spirits or rubbing alcohol into the shoes at the heels and let it soak in. Wear the shoes while still wet.

- Where a shoe pinches over a toe or joint, press a very hot damp cloth over the spot, and leave it for a few minutes so that it expands and softens the leather.

leather bags and luggage

There are different methods of cleaning for different types of leather.

- Brush suede with a suede brush or an off-cut of suede.

- Finish leather with wax polish to keep it supple.

- Smear petroleum jelly on patent leather and then buff it dry.

It's also a good idea to check the manufacturer's care label as both leather and suede are available in two different types — one can be washed but the other should be dry-cleaned.

Luggage deodorizer

To rid a trunk or suitcase of a musty smell, place an open tin containing clay-type cat litter inside. Close the trunk or case lid and leave it overnight.

A new lease on life

To renovate an old leather bag, use a warm, not hot, solution of washing soda to remove grease and dirt. Apply the solution with a soft rag or a brush if the bag is very dirty. Oxalic acid, used after the soda, may remove stains. After cleaning, wash with lukewarm water, place in a warm spot to dry and, once dry, treat with a wax polish.

Don't use soaps or detergent to clean leather as they can cause the leather surface to break down, and will sometimes remove colour.

Substances such as bone,
coral, ivory and shell may
absorb water and soap
during cleaning.

jewellery

A little bit of knowledge goes a long way when
looking after jewels.

- Do store precious jewels in padded boxes or bags
  to protect against sunlight, dust and humidity.

- Do avoid extremes of temperature.

- Do keep precious pieces separate from each other
  to protect against scratches and tangles.

- Do take extra care cleaning jewellery with loose
  stones as the dirt may be all that is holding
  them in place.

- Do take the gentler options first when cleaning,
  starting with a soft brush such as a paint or
  make-up brush on very dusty pieces.

- Do take valuable pieces to professionals for
  cleaning if you are in the slightest doubt about
  how to do it yourself.

- Don't spray jewellery accidentally with hairspray
  or perfume as these can dull some surfaces.

- Don't use any substance on jewellery without
  being absolutely sure it is recommended for all
  the materials the jewel is made from. This
  includes water, detergent, ammonia, bicarbonate
  of soda and jewellery cleaning cloths and dips.

Washing dos and don'ts

- Most jewellery can be washed in a mild detergent and water solution, then gently brushed with soft bristles, but don't wash stones set with glue.

- While transparent gems such as diamonds, rubies and sapphires are hard and do not absorb water, opaque gems such as opals and amber should be wiped only with a damp cloth, then patted dry.

- Hard water can leave a chalky residue on precious stones, so clean them with distilled water.

Washing technique for jewels

Once you are sure a jewel can be washed, put water and washing up liquid in a small bowl. *Don't use the sink in case you lose the item.* Hot (not boiling) water is fine for diamonds, otherwise use lukewarm water. Soak for 15 minutes to loosen the dirt, then rinse. With a small soft brush — for instance, a baby's toothbrush or a paintbrush — work detergent into the crevices. Rinse and dry the jewel on a lint-free cloth.

Caring for wooden jewellery

Wipe with a damp chamois cloth and rub well with a little olive oil. Finally, buff with a soft cloth.

Caring for glass beads

Put the glass beads in a plastic bag together with 2 tablespoons bicarbonate of soda and shake. Remove the beads, dust with a soft brush and buff with a damp chamois cloth.

fragrant home

Pomander

A pomander can be hung in a wardrobe for several months.

1 firm-skinned orange
whole cloves
1 teaspoon orris root powder
1 teaspoon ground cinnamon

Wash and dry the orange and stud it evenly all over with the fresh cloves. Place the orris root powder and the cinnamon in a brown paper bag with the orange and shake the bag to coat the orange. Store the bag in a dark place for a month, then remove the orange and brush it free of powder. Tie a ribbon around the orange and finish with a loop at the top.

Herb and lavender potpourri

30 g (1 oz/1 cup) lavender flowers
30 g (1 oz/$1/2$ cup) dried spearmint
30 g (1 oz/$1/2$ cup) dried marjoram
30 g (1 oz/$1/2$ cup) dried oregano flowers
2 tablespoons powdered orris root
2 tablespoons lavender essential oil

Mix the ingredients together well, then place the mixture in a plastic bag for 2 weeks to mature. Shake the bag regularly. Transfer the potpourri to an ornamental bowl.

Lavender bag

The essential oil present in lavender kills germs and is fragrant. Put lavender sachets or bags in a drawer to prevent their contents from becoming musty. This bag will retain its scent longer if you keep the lavender stalks on.

lavender in full bloom
newspaper
muslin or fine cotton
ribbon

Cut the lavender and spread the stems out to dry on newspaper, either in the sun or in another warm place. Cut the fabric to the size and shape required and stitch into a simple bag shape. When the lavender is dry, insert a bunch into the bag so that the stems stick out of the opening. Close the opening with hand stitching or a length of ribbon. Trim the stalks to a length of 5 cm (2 inches).

A pomander will not only scent your clothes, it will also help to deter moths and silverfish.

clever shopping

Making 'green' choices when you shop can have far-reaching benefits.

Packaging

In industrialized countries, around 30 per cent of plastics are used for packaging, which is used once and then thrown away. While some packaging is clearly useful as protection, some is nothing other than wasteful. Put your money where your beliefs are and buy products with less packaging. Let manufacturers know you disapprove of excess packaging by writing to them.

Eco-labelling

As companies vie for the green consumer's spending money, in some countries there are labelling schemes to help you evaluate claims such as 'recyclable' or 'made from recycled materials'. Consumer groups and government fair trading or consumer affairs departments should be able to give you local information.

Ethical purchases

An increasing number of retailers and direct mail companies are offering textiles, crafts and food products made by people in developing countries, often in co-operatives, who will directly benefit from your purchases. In many cases middlemen have been eliminated from the buying and selling chain so that more profit ends up with the makers of the products.

Organic produce

Organic fruit, vegetables, dairy foods and meat are part of a growing market that offers consumers foods produced without pesticides. These days, even supermarkets are waking up to the fact that some consumers like to choose products that are produced and marketed in ways they are comfortable with. Organic meat comes from animals that spend a certain proportion of their time outside in a non-intensive farming fashion. There are also regulations about the drugs and feed they may be given. Most countries have official organic inspection bodies. Look for their labels when buying organic products.

The green shopper brings their own reusable shopping bag and buys products that use minimal packaging.

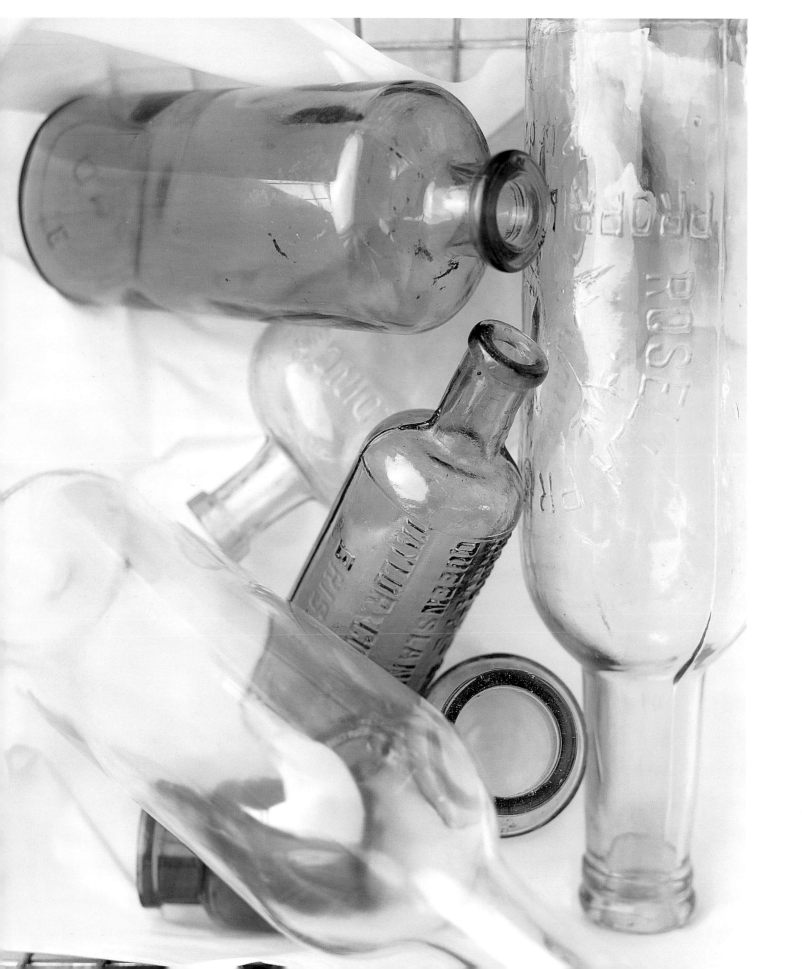

Economy shopping

If you put your mind to it, you need never buy full-price mainstream products again. Numerous outlets offer end-of-line or sales goods at discounted prices — surplus stores, factory outlets, and so on. In addition, generic home brands at major supermarkets and department stores are usually a cheaper option. Buying in bulk, whether at wholesale markets or at the supermarket, can often save money.

recycling

Recycling saves energy in manufacture and also in landfill. It can also save you money. Find out about active recycling schemes in your area. Paper, glass, steel, aluminium and some plastics are commonly collected by local authorities, but tyres, batteries and all sorts of household items can also be reused. You should check with your local authority, as some items may be handled by reverse garbage depots.

Sort and separate your household waste according to what can be done with it: glass in one container, newspapers and other items for recycling in others. Toxic substances, such as old pesticide containers, should be collected by the local authority; vegetable peelings and other biodegradable waste can go into your compost bin.

Find further uses for packaging and other items wherever possible. Cereal and egg boxes can become children's craft supplies, while paper can be reused for telephone message pads, drawing and painting. Old clothing can be cut up for rags or taken to the local charity shop.

Glass can be recycled repeatedly without any loss of quality. Making new products from recycled materials requires reduced amounts of energy and water.

natural garden

respect the natural world,
    infinitely complex and delicate

It is often said that everyone farmed and gardened organically until industrial agriculture was invented. Many of the methods synonymous with modern organic gardening were once mainstream techniques: crop rotation, the use of organic fertilizers like animal manures, non-toxic pest control, composting and mulching. Then things changed drastically garden-wise after the Second World War, when the use of mechanization, synthetic fertilizers and chemical controls became the norm.

By using organic gardening methods we support nature's way of doing things: we become co-operators with nature, not its antagonists. But it is also true that you could garden organically for the most selfish of reasons, for the sheer pleasure of it all. It's easy, it's safe and it yields abundant harvests of plants both ornamental and edible, bursting with life and beauty. Organic gardening blends the discerning use of the latest scientific knowledge with purity and simplicity of approach, methods and materials — soundly based on respect for and understanding of nature. It is a fascinating, satisfying synthesis that *works*.

preparing the soil

Many gardeners put all their efforts into the aboveground parts of their gardens, forgetting what is going on below soil level. Half of every plant lies beneath the soil, feeding and supporting the leaf and flower growth above.

deep cultivation

Any soil cultivation deeper than 20 to 25 cm (8 to 10 inches) can be regarded as a form of deep cultivation. The top 15 cm (6 inches) of the soil is the most biologically active, containing a thriving, organically rich community of beneficial organisms such as bacteria, fungi, insects and worms. These organisms will feed on the organic matter and plant debris in the soil, breaking them down into forms that are available as food for the resident plants. Incorporating organic matter, such as compost or rotted manure, into this zone, or just below it, will improve natural soil fertility, as well as place nutrients close to the roots and improve the texture of the soil. The decomposing organic matter is also very good for retaining moisture close to the plants' roots, something that is particularly beneficial during dry summer conditions. Burying the manure or compost in the bottom of the trench during digging (providing it is not too deep) will encourage deeper-rooting plants.

When you use the deep bed system, the soil structure will be improved to the required depth with one cultivation. From this point onwards, you should keep deep cultivations to a bare minimum so that a natural soil structure develops and remains largely intact. Avoid walking on the soil beds to prevent disturbance or compaction.

A fertile, well-drained soil
will produce healthy,
bountiful vegetables and
flowers that stand a better
chance of surviving the
depredations of pests
and diseases.

deep bed system

Mark one edge of the bed with a planting line. Measure 1.25 m (4 ft) across with a planting board and set up another planting line parallel to the first. Using canes, mark a trench 60 cm (2 ft) wide. Dig it out one spade deep and keep the soil to cover the last trench. Break up the exposed subsoil in the bottom of the trench with a fork, so the roots can penetrate deeply.

Put a 5 to 8 cm (2 to 3 inch) layer of well-rotted manure into the bottom of the trench.

Leaving a cane in the corner of the first trench, measure another 60 cm (2 ft) section with the other cane, so it contains the same amount of soil. Dig the soil and transfer it to the first trench, spreading it to cover the manure.

Add another 5 to 8 cm (2 to 3 inch) layer of manure to the first trench. The bulk of the manure and the loosened soil will raise the bed height. Continue to dig out soil from the second trench and cover the new layer of manure, leaving a deep bed of loose, organically enriched soil in the first trench. Scrape the soil from the bottom of the second trench. Break up the exposed soil. Repeat these steps until the whole plot is cultivated. Use the saved soil to cover the manure in the final trench.

no-dig gardening

Australian Esther Dean first became famous in the late 1970s for a specialized form of sheet composting which became known as the 'no-dig' garden. This technique can be used anywhere, even on the worst and most compacted soils, or directly onto lawn, to provide a flourishing, fertile garden bed.

First provide an edging for the future garden. Then give the soil a good soak. Next, lay overlapping thick layers of newspaper on the ground (cardboard or even old carpet can also be used), and follow this with a layer of lucerne hay, a layer of organic fertilizer, a layer of loose straw, and another thin layer of organic fertilizer.

After watering well, make depressions in the surface and fill them with compost. Well-established seedlings, large seeds, tubers and bulbs can then be planted immediately. If there is sufficient compost available, you can place a layer of compost on top of the garden, rather than creating compost pockets. The contents of your no-dig bed will reduce in height quite rapidly, and will grow excellent crops.

No matter how heart-breaking the original soil is, this method will result in a great improvement in the soil's texture and will encourage the presence of large numbers of earthworms.

Continue to add mulch and layers of organic material to the no-dig bed as it composts down.

fertilizers

Organic fertilizers can encourage soil organisms and lead to better structured soils. Like all fertilizers, they are classified according to their component ratio of nitrogen, potassium and phosphorus (known as the NPK ratio). These are particularly important elements, as they are needed for basic plant growth and function and are used in relatively large quantities.

- Nitrogen is responsible for leaf growth, but too much nitrogen can cause floppy growth and poor flowers.

- Potassium maintains the rigidity of plants and is important in promoting flowering.

- Phosphorus is vital for strong roots and stems.

The 'big three' are all highly soluble, so if you apply excess fertilizer it washes straight into waterways and eventually causes algal growth. It is particularly important to remember this when you are feeding your lawn.

basic steps to feeding

Always fertilize when the soil is moist and water thoroughly after you have completed the application. Never apply fertilizer to dry soil, as there is a good chance this will burn the plant roots.

If in doubt, apply fertilizer at half-strength, twice as often.

Plants don't use much food in winter, so don't bother feeding then. Spring, summer and autumn feeds are generally better value.

Organic fertilizers derive from natural wastes, such as rotted-down leaves and plants, or animal manures.

mineral fertilizers

Even if you feed your soil with good-quality compost, it may still lack mineral nutrients if the original soil is poor. A soil test will show up any deficiencies, which you can fix by the use of rock dusts and other additives. *Calcium* is best supplied by dolomite derived from dolomitic limestone (this form also supplies magnesium to the soil). Ground limestone can be used as an alternative. *Phosphate* is usually supplied as the dust of natural rock phosphate. *Potassium* can be added to the compost heap in the form of wood ashes. *Sulphur* comes in the powdered form 'flowers of sulphur', which is mined from volcanic deposits.

liquid fertilizers

Liquid fertilizers give a rapid growth boost to plants that have been stressed or are under insect attack. Perhaps the finest organic liquid fertilizer is seaweed fertilizer. Some commercially available products incorporate fermented fish and help to stimulate not only plant growth but also desirable microbial activity. Plants with black spot and other fungal diseases respond rapidly after a spray of these substances, diluted as recommended.

Note: Most liquid fertilizers have a rather memorable odour.

Seaweed fertilizer

Visit your nearest beach and fill a sack with seaweed that has been washed up onto the shore. Tie the sack and immerse it in a large container of water. Leave it there for 7 to 14 days. This will make an invaluable, although strong smelling, liquid garden feed when sprayed or poured on to your plants.

Seaweed is a natural source of trace elements, stimulating root growth and improving nutrient uptake.

Manure soup

Follow the same process as for seaweed fertilizer, using a bag of manure. The resultant 'soup' can have a somewhat evil odour initially, but it's an absolute treat for organically grown produce.

Weed fertilizer

Replace nitrogen in the garden by making liquid fertilizer out of weeds. Almost fill a plastic garbage bin with weeds. Cover the weeds with water and replace the lid. Leave until the weeds have broken down. In warm weather this may only take a few weeks. Dilute the fertilizer 10:1 with water and pour it onto the garden.

Blackjack

Blackjack is an excellent, nutritious plant 'pick-me-up', very useful during flowering or fruiting periods.

You will first need a quantity of animal manure that has been well rotted down. Add some soot (which provides nitrogen) and wood ash (good for potassium) to the manure. Put the mixture into a plastic-net bag. Seal the bag carefully and suspend it in a barrel of rainwater. Leave it in position for several weeks.

Once the solution is ready, decant it as required into a watering can, diluting it to the colour of weak tea, and apply it to your plants.

leafmould

Leafmould makes an excellent soil conditioner, but also has low levels of nutrients (0.4 per cent nitrogen, 0.2 per cent phosphate and 0.3 per cent potassium), and is usually slightly acidic.

In nature leafmould is a material that slowly forms beneath trees over many years, so making your own is a long-term project. The leaves can take up to 2 years to decay into a dark, compost-like material.

Rake up the fallen leaves into heaps. Alternatively, run a lawnmower over the leaves with the grass collecting box on. This will not only gather up most of the leaves, but it will also chop them up, accelerating their decay. The best time to gather the leaves is just after it has rained, when the leaves are moist; but they can also be collected dry and dampened later. Make sure you remove any foreign material, such as plastic wrappers.

Collect the leaves and place them in either plastic bags or black heavy duty garbage bags. The latter are better as they block out most of the light and encourage fungal activity. To every 30 cm (1 ft) layer of leaves add a small amount of organic fertilizer, such as dried, pelleted chicken manure or a measure of organic nitrogenous fertilizer such as sulphate of ammonia (which contains 16 to 21 per cent nitrogen).

When the bag is almost full, place it in the position where it is to be left while its contents decompose, and water it thoroughly so that the contents are soaking wet.

Over a period of about 2 years, the leaves will decompose and settle in the bag. These leaves will be pressed tightly together, with some remaining almost whole and others disintegrating completely. When the leaves are ready for use, the bag can be split open and the leaves used as mulch or soil conditioner.

compost

Nature constantly makes compost, that almost-magical, soil-like material which results from the decomposition of organic material and forms the basis of the organic garden. There are several ways to produce good compost, but all techniques fall into one of two categories, depending on whether they rely on bacteria which are aerobic (oxygen users) or anaerobic (non-oxygen users) to break down the raw materials. *Aerobic composting* is very efficient. It generates sweet, nutty-smelling compost rapidly: the compost can reach temperatures of 70°C (138°F), hot enough to sterilize weed seeds and kill many disease organisms. *Anaerobic composting* is much slower and takes place at cooler temperatures. This method is prone to produce undesirable odours, and depends on very ancient and less efficient species of bacteria.

what goes in the compost heap?

Some people say that anything that has lived before can be put into a compost heap. The following items can be included: vegetable and fruit peelings; crushed eggshells; tea leaves and coffee grounds; weeds; soft and hardwood prunings; grass and leaves; seaweed; waste paper and cardboard (individually crumple sheets before use so that they increase the aeration of the heap); sawdust and wood shavings; spoiled hay, alfalfa (lucerne) and straw; and animal manure. When adding manures, note that a mixture of manures works best. Poultry manures are very high in nitrogen and will allow the compost heap to reach high temperatures. Horse and cow manures are a more even mixture of nitrogen and carbon, and while of good quality will not create such high temperatures.

Caution required

Take care when using spent mushroom compost. Mushrooms are often grown with high levels of insecticidal chemicals; the compost is often almost exhausted of nutrients having grown successive flushes of mushrooms; and the pH of spent mushroom compost can be strongly alkaline and unsuitable for acid-loving plants. It is a good idea to recompost it and leave it for a few months before using.

Fish and chicken bones, meat scraps, cheese and other protein-containing material will almost inevitably attract unwanted attention from hungry animals (including rats) and should be excluded from the pile.

Never compost materials that may have been sprayed with chemicals, and never place severely infected prunings or plants in the compost heap.

Crushed eggshells added to the
compost heap will reduce the acid
levels of the carbon-rich material
and stop it becoming too wet.

the right conditions for composting

Compost heaps are often hidden in dank corners, behind shrubberies and in the
dripline of branches. It's far better to place them in a warm, wind-protected
situation (but not an overly hot one, which would dry the heap unnecessarily).

Size matters if a compost heap is to heat effectively. Little compost heaps
never reach the desired temperature, so save up materials to make a larger heap.
Ideally the heap should be at least 1 m (3 ft) wide, long and high — larger is
even better. To stop a big heap falling over as it decomposes, and to make it
less vulnerable to pets and children, enclose the pile within a compost bin.

Well-made compost heaps are moist and hold water like a squeezed sponge. If the
pile has been made off the ground, any excess water, which would displace air
from the pile, cannot accumulate. If your compost heap gets too wet, it will
not heat: as a result, it will become anaerobic and start to smell unpleasant.

The greater the surface area of material exposed to the activities of composting
micro-organisms, the more rapidly they will be able to convert a compost heap to
usable compost. A shredder can be invaluable to help reduce woody materials to
chips, which have an infinitely greater surface area than the original stems and
branches. Alternatively, chop larger pieces of prunings and clippings with a
spade. And get the hammer out to break up the tough stems of corn stalks and
cobs, broccoli and Brussels sprouts stems, and other tough vegetable remains.
Crush eggshells before adding them to the kitchen vegetable scraps container.

when is the compost ready?

Aerobic composting is much faster in warm weather. The time taken in hot, humid
summer areas may be as short as 14 to 20 days, and in cooler areas 2 months.
You know the composting process is finished when the pile has completely cooled
and the content has the appearance and texture of rich brown, crumbly earth. It
is not unusual to find the remains of a few tough, fibrous ingredients like
corn cobs. If you want, sift the compost by tossing it in forkfuls through a
panel of wire netting. Then add any uncomposted remains to a new pile.

Protect the top of the finished pile of compost with a plastic sheet or
tarpaulin. Otherwise, if the pile is left exposed to the rain, the nutrients
will be leached out.

When using tree or shrub prunings in the compost heap, shred them or cut them down so they are as small as possible.

engineering a compost heap

A good compost heap is like a well-laid fire in a fireplace, allowing oxygen to be drawn up and vented through the heap. The easiest way to achieve this is to build the compost pile on top of an open 'mattress' of thick, branched sticks. Larger tree and woody shrub prunings are ideal. Alternatively, you can use a section of heavy-gauge wire lattice balanced over bricks placed at each corner.

To provide venting inside the heap, stick four or more stakes, depending on the intended size of the heap, vertically into the pile after the first layers have been established. Continue to build the pile around the stakes, and when it is complete, wriggle the stakes loose and carefully pull them out. This creates 'chimneys' through the pile that will draw air upwards. Material like freshly mown grass (which is moist and tends to compact in the pile) should be mixed with drier materials like sawdust or straw.

Some gardeners dismantle the heap when it begins to cool down, and invert the pile by forking the outside to the inside and vice versa. This fluffs up the pile and makes air readily available. Material turned into the centre is exposed to the warmer, moister heart of the pile where the greatest microbial activity occurs. The pile will heat up for a second time. This step is not essential: if you leave the pile as is, complete composting will simply happen more slowly.

constructing a compost bin

Simple wire netting compost bins are ideal. These are easy to make and to dismantle, allow excellent access for air, and are not themselves composted. They are cheap to build and the materials can be readily recycled. Tall, vertical, metal star posts or wooden posts are embedded in the soil to make the corners, wire netting forms the sides, and wire ties can be used for the construction.

Some gardeners construct the walls of the heap with bales of straw arranged with small gaps between. Less air is accessible but the walls themselves contribute to the compost and are easily dismantled when composting is completed. The partly composted hay bale walls can become the foundation of the next heap. In cool climate areas, hay bale walls help to insulate the pile from heat loss.

Never construct a compost heap or bin around a tree. The bark will compost, letting in unwanted disease organisms.

commercial compost bins and tumblers

Commercial bins are popular with some organic gardeners. They don't take up much space and they effectively exclude animals — both household pets and vermin. Resembling garbage bins, they do provide some insulation and allow microbes to remain active longer in the season. They can look very neat and tidy, if that is a priority for you.

Compost tumbling systems are usually fairly low volume but will still handle all the household scraps generated by a small family. Under ideal conditions they will convert organic material to compost in as short a period of time as 3 weeks — approximately the time of a well-constructed aerobic compost heap.

worm farms

Worm farming can be an excellent way of dealing with modest amounts of household scraps, and at the same time producing quantities of excellent natural fertilizer.

Commercial worm farms are available and it's quite a simple matter to follow the instructions in order to assemble them. But a worm farm can also be made from a simple bin or box, vented with air holes around the side and about 5 cm (2 inches) below the rim. You should also make a row of drainage holes.

Keep in mind that the worms are active only between the temperature range of 12°C to 25°C (50°F to 77°F). Make sure the bin is moved into a warm area and insulated throughout winter in colder regions, and placed out of the sun and in a cool area during hot summers.

Only species of worms adapted to living in decomposing organic matter are suited to worm farming, such as *Lumbricus rubellus* and the red wiggler worm or Brandling worm *Eisenia foetida*. These are available at some garden centres and can also be mail-ordered.

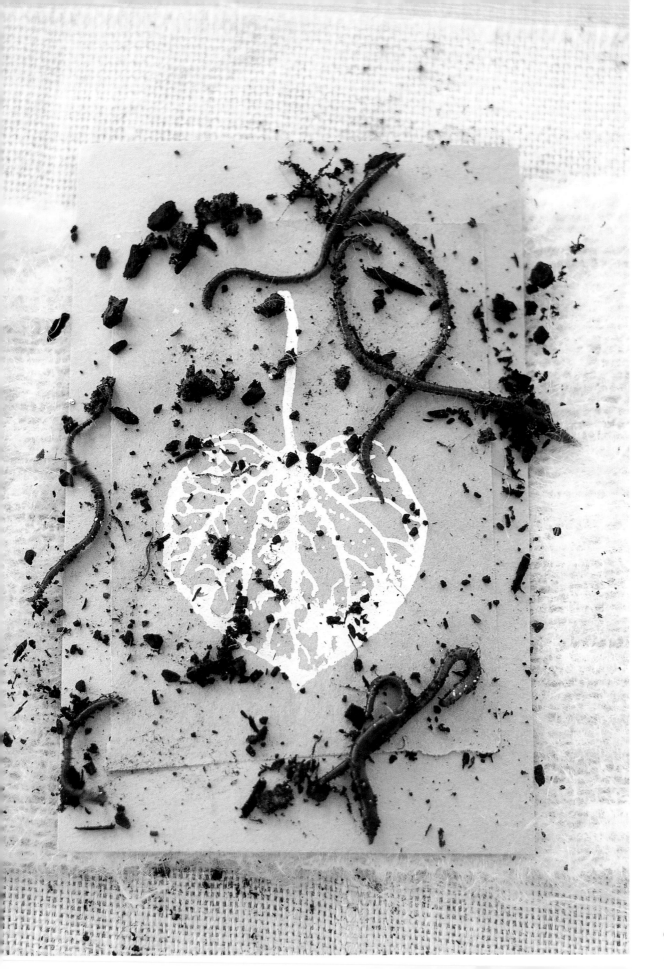

trench composting

Trench composting can be an ideal answer for those who have larger gardens and for the more patient among us who are willing to wait for results. Trench composting is an anaerobic technique, so the compost materials will not reach the temperatures needed to kill weed seed and pathogens.

As this is a cool process, earthworms will actively contribute to turning the organic matter through the soil. You could add commercially obtained earthworms to the trench, but populations of native worms will be attracted to the organic waste and multiply rapidly of their own accord.

Woody materials such as prunings will decay quicker if they are shredded before being buried, and a small amount of nitrogenous fertilizer may also need to be incorporated to speed up the whole process.

In the late summer or early autumn, mark out the area which is to be dug over in a series of trenches and mark the lines of the parallel trenches.

Dig out a single trench about 30 cm (1 ft) deep, and move the soil from the trench to the end of the plot, which will be the very last section to be trenched.

As they become available, gradually fill the trench with plant debris, vegetable scraps and kitchen waste.

Dig out a second trench in a similar way to the first one. Cover each additional layer of material in the first trench with the soil dug from the second parallel trench.

After the first trench is full, start filling the second trench by creating a third trench. Each completed trench will gradually settle over a month or two as the plant material decomposes.

green manures

Green manure crops are planted to improve the quality of the soil and enhance plant growth. They are usually nitrogen-fixing plants, often with strong, deep root systems that help to break up compacted soil and draw nutrients up to the soil surface where they are available to future plantings.

Alfalfa (lucerne) *Medicago sativa*

Alfalfa is a perennial plant that prefers neutral to alkaline, well-drained soils. Plant from spring to mid-summer. Alfalfa has nitrogen-fixing qualities and is exceptionally deep-rooted.

Buckwheat *Fagopyrum esculentum*

This annual is good on poor soils and will improve soil structure as well as attract pollinating bees. Plant from spring to mid-summer.

Crimson clover *Trifolium pratense*

This plant lasts 2 to 5 months and will overwinter. It prefers sandy loam soils and should be planted in spring to late summer. A nitrogen-fixing plant, it also attracts pollinating bees.

Daikon (white radish) *Raphanus sativus*

Lasting 3 to 5 months, daikon is tolerant of most soil conditions. It should be planted in spring. Its flowers attract beneficial insects, while its huge roots break up soil for water penetration.

Lupin *Lupinus angustifolia*

Lasting 3 to 5 months, lupin can be grown in acid, sandy to sandy loam soils. Plant in spring to mid-summer. It is a nitrogen-fixing plant.

Scorpion weed (phacelia) *Phacelia tanecetifolia*

Lasting 2 to 4 months, this plant can overwinter. It has a wide soil tolerance and will improve soil structure as well as attract beneficial insects. Plant from spring to summer.

Rye *Elymus* sp.

This plant has a wide tolerance of soils and will overwinter. Plant in early to late autumn. Its fibrous root system improves soil structure.

mulch

It is impossible to overestimate the value of mulches in the garden. They are invaluable in conserving soil moisture so that plants are less prone to water stress, and therefore less prone to reduced yields and to pest and disease attack. Mulches also minimize erosion, help maintain an even soil temperature, and reduce soil splash on plants, which can spread soil-borne diseases.

how mulch works

Mulch should be spread to provide a blanket layer over your soil that is normally about 10 cm (4 inches) thick. This layer regulates soil temperature by keeping the plant roots cool in summer and warm in winter. It also conserves moisture and cuts down on watering requirements by reducing evaporation from the soil surface and increasing water penetration. Mulch also controls weeds by preventing weed seeds from germinating.

organic versus inorganic mulch

Mulches come in organic and inorganic forms. *Organic mulches* have additional benefits. As they break down they provide organic matter to the soil, improve soil structure and encourage beneficial micro-organism activity. Organic mulches include leaf mulch, pine bark, red gum chips, alfalfa (lucerne), straw, newspaper, compost, rice husks and sugar cane. An effective mulch should not be dislodged by wind and rain and should have a loose enough structure to allow water to soak through easily.

Some mulches — for example, alfalfa (lucerne), compost and sugar cane — have a high nitrogen content. They improve soil fertility, but rot down quickly and so need to be replaced every few months. Never use peat moss as a mulch, as it repels water once it is dry — it is also a non-renewable resource.

Inorganic mulches such as black plastic, weed control mat, scoria and decorative gravels and pebbles are not 'garden friendly'. They add nothing to the soil structure and once they are in place, make it difficult to incorporate soil additives. They tend to raise the soil temperature and some can even stop your soil from breathing.

Living mulch

Many low-growing plants make ideal living mulches. Groundcovers such as ajuga (blue bugle flower) planted into mulched soil are excellent for excluding weeds in ornamental gardens. Vinca (blue-flowered periwinkle) is ideal for dry shade areas and is easily controlled at a suitable height with a string trimmer used twice a year. Any of the prostrate-growing plants are effective in reducing weeds.

However, weeds may not always be a nuisance. They can, in fact, be helpful. A carpet of weeds can act as a protective blanket, another 'living mulch' of sorts.

Many gardeners make their lives a misery worrying about weeds, but organic gardeners tend to be a bit more relaxed about 'invasion' by 'unwanted' plants. They know that weeds are essential to the health of the soil. Bare earth is easily eroded by wind, and compacted by heavy rain or foot traffic. It is more easily leached of soluble nutrients and can also lose important gases.

If an area is to remain unplanted for a while, allowing it to become covered in weeds may not be the neatest solution, but it is sound ecologically. The weeds can be slashed just before they begin to flower and left on top of the soil as a green mulch, or dug through to add valuable organic matter.

Apply light layers of mulch that allow air and water to penetrate the soil beneath.

When using lucerne hay as a mulch, ensure there are no seeds. Lucerne is said to provide protection against fungal diseases and root-rot.

applying organic mulch

You should apply any organic mulches at least 10 cm (4 inches) deep. To exclude light from the soil surface and help weed suppression, spread overlapping layers of newspapers before applying the mulch.

Mulches should not be laid until the soil has warmed up after winter. And don't place mulch too close to the trunks of shrubs and trees, as it can cause collar rot.

Fruit trees benefit greatly from a mulch of nitrogen-rich alfalfa (lucerne) hay which can be weighted down with stones. Rake up fallen fruit and leaves at the beginning of winter and compost them. If left as a mulch they will act as a reservoir of disease for the following season.

Sawdust and wood chips are often used as mulch. However, when either raw sawdust or wood chips begin to break down, the responsible soil bacteria require nitrogen for the process. They will rob the soil, and plants that have been mulched with raw sawdust or wood chips will exhibit yellowing due to this temporary withdrawal of nitrogen. Once the process has been completed, the nitrogen is made available to the soil. You can either add a sprinkling of blood and bone to provide the required nitrogen, or partially compost the sawdust or chips in a pile for a month before use. Sprinkle the pile well with liquid seaweed fertilizer to boost the partial composting process. Sawdust and woodchips often contain levels of tannins high enough to inhibit plant growth, and these are also partially leached during composting.

Mulch is invaluable, but it is better not to simply throw masses of garden refuse on the soil. Compost it first. Piles of rubbish will decompose anaerobically into a slimy mess that encourages disease. Of all mulches, none are better than compost (which can be applied when quite roughly textured) and alfalfa (lucerne) hay.

The introduction of a legume crop such as peas or lupins provides free nitrogen fertilizer.

crop rotation

Crop rotation is simple in principle: don't plant the same crop in the same bed for 2 years running. Most importantly, crop rotation avoids the build-up of any crop's particular pests and diseases. It also helps get the best value from your compost. The need to grow vegetables quickly to make them tender means they must have rich soil, and that means the vegetable garden will take most of your compost. But not all vegetables need their soil equally enriched. Leafy things such as lettuces and cabbages need it richest; fruit crops such as tomatoes, peas and capsicums (peppers) are not so greedy; root crops are less greedy still.

So, plant a root crop such as turnips or carrots where you've just had a leaf crop such as lettuce. Plant onions prior to a crop of tomatoes, and peas and beans after cabbages and silverbeet (Swiss chard). Avoid planting anything in the same family season after season in the same bed: do not plant tomatoes after potatoes, or cabbage after broccoli. Keep records of what you plant and where. This list will help you to identify the vegetable groups.

- Cabbage, Chinese spinach (amaranth), cauliflower, broccoli, Brussels sprouts, kohlrabi, radish, swede (rutabaga), turnip, mustard greens

- Tomato, pepper (capsicum), chillies, eggplant (aubergine), potato

- Pea, broad (fava) bean, dwarf bean, climbing bean

- Pumpkin, squash, melon, cucumber, zucchini (courgette), marrow

- Carrot, parsnip, celery, parsley

- Swiss chard (silverbeet), beetroot, spinach

- Lettuce, globe and Jerusalem artichoke

- Onion, garlic, shallot, leek, chives

companion planting

Companion planting is an effective way to protect plants from the unwanted attentions of pests and diseases. A number of plants give off strong odours that can confuse the olfactory senses of pests. Many herbs are useful for this purpose, including basil, lavender, rosemary, rue, thyme, chives and garlic. For them to be effective, you should plant them throughout the garden.

However, companion planting is much wider in scope than just confusing an insect's sense of smell. In its broadest sense, it includes any plant that is beneficial in some way to another plant. Think of companion plants as being 'good neighbours'.

It may be a simple matter of one plant shading another or modifying the humidity. One plant's roots might aerate the soil, or help drain excess water. And some plants can protect others as much as themselves by virtue of defence mechanisms such as thorns and stinging hairs, or by producing compounds that are poisonous to insect pests. Other plants offer benefits to their neighbours by attracting or housing desirable insect predators, or by exuding odours that attract insect pollinators.

marvellous marigolds

Nematodes (eelworms) cause reduced growth, low yields and wilting in a variety of vegetable crops. When infected plant roots are dug up, they appear to have tiny gall-like growths over the surface (gardeners often mistakenly attribute these effects to drought or poor soil fertility). However, nature has produced a powerful nematicide, a biomolecule produced to some degree in the roots of all species of marigolds (*Tagetes* spp.). The dwarf French marigold (*Tagetes patula*) is particularly useful for barrier plantings around gardens. If the brilliant golden or orange flowers upset your garden colour scheme, shear the heads off.

trap plants

Some plants are particularly attractive to pests — because of their colour, smell or taste — and thereby protect other plants from attack. Bright yellow nasturtiums attract aphids away from cabbages. Zinnias have long been used as trap plants to lure Japanese beetle. Dill is traditionally used to lure green tomato caterpillar. In themselves, trap plants are not sufficient protection for your garden, but they do contribute towards maintaining healthy crops.

Nasturtiums deter the Brassica pests blackfly and whitefly. They can also be harvested for salads.

The delicate flowers of Queen Anne's
lace (*Daucus carota*) attract beneficial
insects, such as bees.

good companions

These plants have long been regarded as good companions:

- Basil with tomatoes, asparagus, beans, grapes, apricots and fuchsias
- Beans with potatoes and sweet corn
- Chives with carrots, cucumbers and tomatoes
- Cucumbers with potatoes
- French marigolds (*Tagetes* spp.) with tomatoes, roses, potatoes and beans
- Hyssop with cabbages and grapes
- Leeks with celery
- Lettuce with carrots, onions and strawberries
- Melons with sweet corn
- Mint with cabbages and other brassicas and peas
- Nasturtiums with cucumbers, zucchini (courgettes), squash and apple trees
- Onions with carrots, kohlrabi and turnips
- Peas with carrots

bad companions

Some bad combinations to watch out for include:

- Apples with potatoes
- Beans with garlic
- Cabbages with strawberries
- Gladioli with strawberries, beans and peas
- Sunflowers with any vegetable but squash
- Wormwood with almost everything

Choose seeds from healthy and
productive plants that have
that variety's typical — and
desirable — characteristics.
Don't just harvest from the
first plant to go to seed.

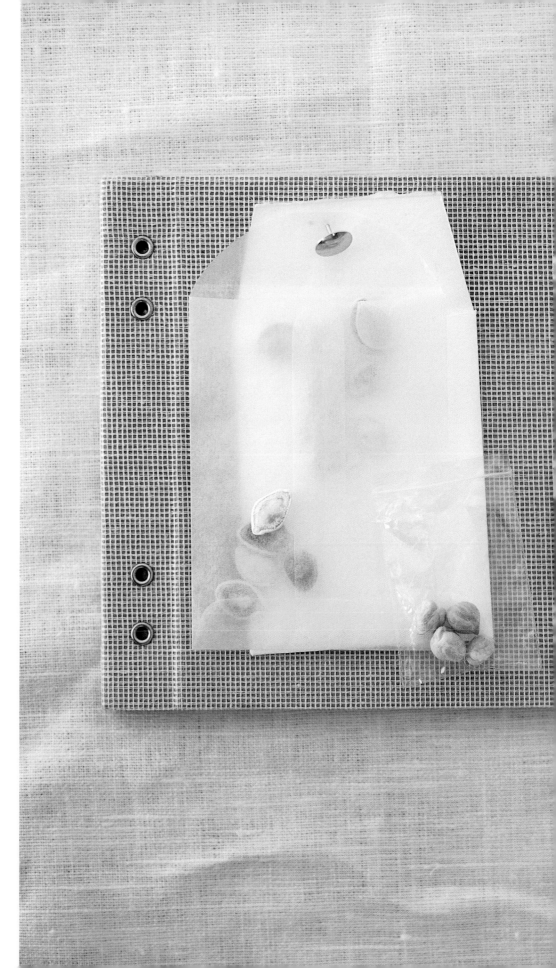

seed saving

Various techniques are used for seed collection, depending on the growing and seeding habits of the plant.

Plants in the Apiaceae family (previously Umbelliferae), including celery, carrots, celeriac and parsnips, form lacy heads of tiny flowers followed by progressively ripening seed. To prevent much of the seed spilling on the ground, pick the entire head once the most mature seeds are fully sized and brown. Then, place in a paper bag, tie with string and suspend upside down in a well-ventilated area out of direct sunlight. The seed will all fall into the bag. Clean the seed, looking for any insects, place in an airtight container, label (including a date), and store in a dry, cool place.

Apiaceae seed is generally viable for only 1 year, so you should plant a crop every year. Seed of the onion family, Alliaceae, is treated in the same way.

Capsicum (pepper) and chilli pepper seed can be removed from the fully ripened fruit and dried for storage. It is a good idea to wear gloves for this process, as the chilli oils are hot.

Peas and beans should be left growing in the vegetable patch until the pods are dry. Pull the bushes from the ground at the end of the season and thresh them on a sheet to release the seed.

Brassicas too should be left in the garden. They will explode their seed as soon as they are fully dried, so enclose the heads in large paper bags to capture as many seeds as possible.

pests and diseases

The more healthy a plant is, the more resistant it will be to pest and disease attack. Insect pests are attracted to the weakest, most stressed plants in a crop. Improving soil structure and fertility with organic matter makes for healthier soil and this in turn makes for strong, healthy crops — and minimal crop damage. Organic gardening techniques also produce plants that grow steadily rather than rapidly as is the case with chemical fertilizers. The plants therefore do not become soft and sappy and prone to attack.

Compost and mulches retain soil moisture so that plants are not water-stressed on hot days. It is also important to water in response to the needs of the plant. Plants that are regularly wilted and stressed, as well as ones that are overwatered, are much more susceptible to attack by insects.

A plant in the wrong position is also liable to be attacked. We all succumb to the temptation to buy and plant things that are marginal for our climate, do poorly in our type of soil, or need conditions we cannot offer like wind protection or an open sunny site. But plants that survive these impulsive gardener moments will always remain more vulnerable to pests and diseases.

And it pays to plant crops at the optimal time. Plants may well survive being planted too early or too late, but they will never thrive as they should.

You can also choose cultivars that have been selected for resistance. Many strains of vegetables are able to resist attack to some degree. If you garden in an area regularly affected by a particular pest or disease, some research and careful selection will reward you with stronger, less susceptible plants. Neighbourhood plant nurseries can be invaluable sources of good advice.

good garden housekeeping

Many diseases and plant pests can be eliminated from the garden simply by good housekeeping practices. All garden wastes, including spent crops, should be composted. If material is infected, it should be placed in the centre of the compost heap where the high temperatures reached will kill all spores. Infected woody prunings, however, should be burned, if possible.

Viral diseases are passed on mainly by sap-sucking insects. As soon as a virus-infected plant is detected, it should be removed and added to an activated compost heap. Or better yet, burn any diseased plant material if you can.

The organic gardener accepts imperfect and blemished plants, but sometimes intervention is necessary. When deciding if and how to act, inspect damaged plants closely. Monitor your garden for early signs of problems, and aim for prevention rather than cure.

Make a trap for snails by placing cabbage leaves on the ground overnight. Snails will be attracted to the juicy leaves, and you can collect and destroy them next morning.

Dealing with garden problems as they arise will ensure fewer pests and disease problems. A thorough clean-up at the end of summer or early in autumn can do much to prevent pests and diseases in the next growing season. Digging the garden over at this stage not only aerates the soil but can also expose overwintering larvae of various pests. Make sure that no vegetables are left on the ground. Any mummified vegetables should be burned if possible. After pruning deciduous trees, check for the presence of borer and destroy any you find by poking a wire into any holes. Use a wire brush to remove any loose bark, which often shelters overwintering pests.

don't get mad, get clever

Combating slugs and snails

If your garden is small enough, picking off slugs and snails every night will reduce the damage done by them. Just squash them under your shoe or boot. Beer in a saucer is another effective trap — you will need to empty it every morning. Soft mollusc bodies don't like crawling over scratchy materials such as crushed eggshells, slaked lime, salt or sawdust. Snails and slugs will also not cross copper because it gives out a weak electrical charge; trees can be protected with a band of copper, but allow room for a little growth.

Barriers and traps

Among the most useful advances in recent times has been the development of finely woven, transparent cloths to protect vegetables and fruit trees. These are woven to allow water and maximum light and air through while excluding insect pests. Floating row covers are ideal for the vegetable garden.

Other relatively newly developed barriers are sticky, non-drying glues that trap insects migrating up the stem or trunk. The glue is placed on a paper collar around the base of the plant. A simple non-sticky collar can be made from a cardboard cup with the base cut out. Placed around the base of a seedling, this is sufficient to protect it from cutworm damage.

In some areas, carrot fly is a real problem. But the female fly needs to hover low over the crop in order to detect the odour of carrots. Erecting a simple, temporary barrier fence of hessian (burlap) around the row will force the female fly to hover too high to detect the scent.

to spray or not to spray?

Chemical sprays will upset the ecological balance in your garden and result in a cascade of further problems. Beneficial predators will also be wiped out so that pests previously under control may take over. Instead, try one of the organically acceptable sprays, which include garlic spray, wormwood spray, pepper spray, pyrethrum spray, and the fungicides lime sulphur and Bordeaux mixture. The latter contains copper and hydrated lime and is used on potato blight and other fungal problems. Bordeaux spray should be restricted in use as it builds up levels of copper in the soil. And lime sulphur should only be applied when a plant is dormant.

Pyrethrum is derived from the white flower heads of an African daisy species and is useful against aphids including greenfly and whitefly. But note that pyrethrum can also kill useful insects. Soapy water spray made from soft soap (not detergent) is useful against aphids.

Insect-repelling 'teas'

'Teas' can be made with plant material known to be insect repellent. Chop the leaves roughly, cover with water in an enamel saucepan and simmer for 15 minutes. Strained and cooled, the resulting tea can be used as a spray. Try eucalyptus, wormwood, southernwood, black sage (*Salvia mellifera*) and equisetum. Ensure no one is tempted to drink these teas, and that they are well labelled.

Homemade organic sprays

A number of old-fashioned homemade sprays are remarkably effective in helping to restore balance in the garden when a pest reaches the nuisance level. While these sprays may temporarily clear the kitchen with their odour, they certainly won't endanger you or your family.

Less toxic pest control methods may need more frequent application. Controlling aphids with garlic spray may mean spraying every 3 days.

Quassia tea

100 g (3½ oz) *Picrasma quassioides* wood
   chips
3.5 litres (3½ qt) water

Very gently simmer the chips in the water
for 2 hours. Strain and cool. Dilute the
tea 1:5 with water for spraying pests.

Garlic spray

This makes a good all-purpose spray. The
garlic works on sucking insects such as
aphids on roses, while the chilli works on
chewing insects such as caterpillars.

1 bulb garlic, broken into cloves
250 ml (9 fl oz/1 cup) water
3 chilli pepper pods
1 tablespoon soft soap or grated soap
   flakes

Place all ingredients except the soft soap
in a blender. Filter through a double
layer of cloth or coffee filter paper.

To use, dilute the liquid with 3.75 litres
(8¼ pt) of water and add the soft soap.
Apply with a clean spray gun.

Onion spray

155 g (5½ oz) roughly chopped onions
250 ml (9 fl oz/1 cup) water
5 feverfew leaves (for extra impact)

Prepare, dilute and use as for the
Garlic spray.

waterwise gardening

Around a third of household water is used in the garden, so any steps you take outside to reduce your usage will have a big impact on your overall water consumption. Here are some ways to avoid excessive use of the hose.

- Native plants. Native plants, whether they are shrubs or trees, tend to need less water than exotics.

- Windbreaks. Plant windbreaks to reduce the drying effect of the wind, and cover bare soil by planting or by placing pavers and boulders. Bare soil heats up in the sun, drawing moisture from underneath to the top, where it evaporates.

- Mulch. Mulching garden beds will help stop the soil from drying out and reduce your watering needs.

- Maximize. Water efficiently: avoid using a hose in the middle of the day, when water will evaporate quickly; fit timers to sprinklers; choose sprinklers with big drops, not fine mists that are blown away; consider installing a fixed watering system with drip-feeders that deliver water directly to plant roots and eliminate wasteful run-off; group plants by their watering needs; water slowly to allow water to soak into the soil.

- Train. Give your garden a good soak once a week rather than a daily drip. This 'trains' your plants, encouraging bigger and deeper root systems and thus hardier plants. Frequent watering results in shallow roots, which aren't so good at seeking out water.

- Sweep. Use a broom, not a hose, to clear paths.

- Green grass. Lawns are thirsty and mowing them uses polluting fuel (unless you use a push mower). Consider grasses that need less water and different kinds of ground cover altogether. If you stick with a lawn, keep it on the long side, about 2 cm ($^3/_4$ inch), to shade the ground. Water no more than twice a week even when it's hot. Where possible, wash the car on the lawn to give the grass a watering.

Organic gardening
methods, such as
the use of mulches
and natural soil
improvers, help to
conserve water.

Rooftops make perfect rainwater catchment areas, but certain roofing materials, for example lead- and tar-based paints and asbestos, can contaminate the water.

rainwater tanks

You can collect rainwater by channelling it from the roof into a tank. Check with your local authority about restrictions on size, height and location of the tank. Also consider what the tank is made from — usually concrete or galvanized iron, fibreglass and recycled plastic; where the tank will be positioned (buried or raised); and what it is lined with (linings may increase the life of the tank and enhance water quality). You will need a non-porous tank interior to discourage algal growth.

grey water

Water used in the house can, in theory, be recycled. The garden is an obvious recipient. How readily recyclable your used water is depends on what you've put into it.

Laundry and kitchen water are more suitable if you use low-phosphate detergents or soaps. Liquid detergent, especially concentrate, is preferable to ones with fillers and softeners. Salts in detergents, particularly powders, cleaners and softeners, may alter the soil pH, making it more acid or alkaline. This can affect how the soil drains, especially if grey water is adding to existing salts in the soil. Signs of problems include water sitting on the soil rather than soaking in, 'burnt' leaves and stunted plants.

Boron in some laundry products may poison plants. Bleaches and disinfectants can poison plants and destroy useful micro-organisms in the soil. Nutrients such as phosphorus and nitrogen in grey water may be beneficial for some plants but harmful to others; be selective about where water is used.

Oil and grease in kitchen water may form a waterproof barrier over the soil, preventing water seeping in.

gardening by the moon

Sometimes observations made over many centuries and verified by long human experience are simply not explainable by science. The fact is that many gardeners and farmers — practical people who rapidly drop any idea that does not yield results — rely on gardening by the moon.

At least part of the influence of the moon is related to its gravitational effect. Both the sun and the moon cause gravitational effects on the earth, but the influence of the very much smaller moon is greatest due to its close proximity. Early research has also shown that during the increasing or waxing moon, seeds and transplants take up water more readily than those sown in the decreasing or waxing time of the moon. As a result, leafy above-ground vegetables appear to be favoured by the waxing moon, while below-ground plants requiring less water are favoured by the waning moon.

the rules of planting by the moon

New moon waxing to full moon

Now is the time to plant the seed and seedlings of annuals that provide their harvest above ground. Garlic and grain crops also do well if planted during this waxing period. It is a good time to sow or lay turf, plant roses, graft new flowering and fruit trees, and plant ornamental flowers.

Waning moon

This is considered the best time to plant biennial and perennial plants, and root and bulb crops.

From the new moon to the first quarter

This is the best time for planting leafy annuals, particularly those that do not bear their seeds in fruits, for example spinach, cabbage, celery, lettuce and endive.

From the quarter moon to the full moon

Plant species that have a vining habit or bear their seeds in fruits, for example beans, peas, cucumbers, squash, capsicum (peppers), eggplant (aubergine), tomatoes and melons.

The moon affects the earth's atmosphere in ways that can benefit the observant gardener: for example, there is a strong possibility of heavy rain following a planting made during a full or new moon.

Many gardeners believe that plants, drawing water from the soil and consisting mainly of fluid, are influenced by the moon, just like the ocean's tides.

From the full moon to the third quarter

This is the time to plant all root and bulb crops, as well as perennials and biennials. This includes potatoes, onions, rhubarb crowns, grapes and berries. Trees and fruit trees should also be planted.

The fourth quarter

Plant nothing.

the moon and gardening activities

Many garden activities are also governed by the phases of the moon.

Waxing moon

All propagation is favoured, including taking cuttings, and budding and grafting. Once cuttings are rooted, they should also be potted during a waxing moon. This is also the best time to repot houseplants.

Watering the garden is considered to be most effective during the waxing phase of the moon, although seasonal conditions, current weather and soil type obviously should also determine the need for water in the garden.

Due to the higher water content of plants in the waxing phase, fresh fruits are at their juiciest and salads at their crispest if harvested in this phase. Grapes are preferably harvested just before the full moon.

Waning moon

Activities associated with the waning phase of the moon include dividing perennials (in late autumn to winter), making and turning compost heaps, mulching, killing weeds and pruning. It's also the time for harvesting and drying herbs for ease of drying and high potency, harvesting and drying flowers, and ploughing or hoeing.

Fourth quarter

Many gardeners consider this the best time to use organic sprays on fruit trees.

Index